The Shape of Nature

Satyan L. Devadoss, Ph.D.

PUBLISHED BY:

THE GREAT COURSES
Corporate Headquarters
4840 Westfields Boulevard, Suite 500
Chantilly, Virginia 20151-2299
Phone: 1-800-832-2412
Fax: 703-378-3819
www.thegreatcourses.com

Satyan L. Devadoss, Ph.D.

Associate Professor of Mathematics
Williams College

Professor Satyan L. Devadoss was born in Madurai, India, on November 26, 1973, and moved to the United States in 1981. He graduated from North Central College in 3 years (in 1993, at the age of 19), during which he won the Outstanding Student of Mathematics award each year and was class valedictorian. He then went to graduate school at Johns Hopkins University, where he was the first recipient of the William Kelso Morrill Award (1995) for excellence in teaching mathematics. In 1997, he was awarded the Dean's Teaching Fellowship to design and offer a course of his choosing.

After receiving his Ph.D. in Mathematics in 1999, Professor Devadoss became a Ross Assistant Professor at The Ohio State University, where he created a course about shapes in nature and received the 2001 Freshman Research Seminar Award. In 2002, he was supported by the National Science Foundation (NSF) to attend the International Congress of Mathematicians in Beijing.

In 2002, Professor Devadoss joined the faculty in the Department of Mathematics and Statistics at Williams College. Here, he received NSF grants for his work in topology and computational geometry. In 2008, he was awarded North Central College's Young Alumni Award for excellence in his career and for demonstrating service to the community and the college.

Over his career, Professor Devadoss has designed more than a dozen novel courses in mathematics, computer science, and studio arts. He has organized art-math symposiums and undergraduate conferences and has been invited to give more than 50 lectures. He has supervised 8 undergraduate and graduate student theses and directed 4 NSF Research Experiences for Undergraduates programs at Williams College.

Professor Devadoss has published more than 12 papers, including work on configuration spaces, cartography, polytopes, origami, triangulations, and juggling. He has been invited to be a member of Mathematical Sciences Research Institute (MSRI), a think tank in Berkeley, California. He has also recently cowritten a book on computational geometry with Joe O'Rourke, one of the founders of the field.

Professor Devadoss is happily married (at least from his end), with 3 kids. He lives in Williamstown, Massachusetts. He cherishes his lack of exercise and the joy of eating ice cream. ■

Table of Contents

INTRODUCTION

LECTURE GUIDES

Table of Contents

The Shape of Nature

Scope:

The world around us is filled with intricate and amazing shapes, both seen and unseen. This course will not only provide a masterful guide to the sweep of designs in nature but will also equip us with concrete mathematical tools to tackle cutting-edge problems. We will focus on things of very small scale, learning about string theory and quantum entanglement in physics, about origami folding and how it shows up in protein design, and about knotting of DNA strands and how to untangle them. We will also look at nature from a grand scale, learning about general relativity and the curvature of planets and stars in spacetime and even about the shape of our universe and how we can try to find it. Using vivid visual imagery and motivated by real-world problems, we will explore mathematics with such simple ideas as tying strings and folding sheets of paper to bring us to the forefront of scientific research.

The vision for the course can be seen as a two-stranded braid. One braid focuses on shapes and designs that appear in nature and looks into the fields of biology, chemistry, physics, and more. The other braid provides the mathematical tools and ideas to understand, manipulate, and explore these shapes and designs. Although our motivation will always be linked to the natural world, we will sometimes take mathematical side roads to peek behind the machines that make things work. For example, the simple notions of addition and multiplication will be seen in a fresh perspective as they appear in the world of shapes. The overall structure of the course follows a mathematical perspective, building on simple structures and moving toward deeper complexity. Throughout these lectures, we present numerous unsolved problems in the world of shapes that are easy to state but thus far have resisted the attacks of talented researchers.

The introductory lectures set the stage for the main attractions. We look at the language of shapes as seen through the eyes of geometry and topology, then study the notion of dimension, where we see not only how to break shapes into levels of complexity but how they can interact with their environment.

The course is divided into dimensional settings, starting with one-dimensional knots and links, moving to the geometry of two-dimensional surfaces, and exploring the world of three-dimensional manifolds. The closing lectures go beyond the standard notions of shapes and peer into the unseeable secrets of nature by considering *n*-dimensional objects, such as spaces of phylogenetic evolutionary trees, particle collisions, and string theory. From higher dimensions, we move to fractal dimensions and chaos, a concept undreamed of until a few decades ago. In the penultimate lecture, we explore the intersection of mathematics and the visual arts, looking at a number of great artists and following their works as they influence and are influenced by the scientific world around them. Finally, we take a look back at some of the ideas and results we have seen throughout the course and highlight some mathematical challenges for the 21st century. ■

Understanding Nature

Lecture 1

The most important lesson that we can learn in these entire series of lectures is that shapes determine purpose. Form and function are interrelated.

In this course, we will embark on an amazing journey in the study of shapes, ranging from big, puffy clouds to the patterns on ties and shirts. Why do we care about shapes? Because shapes determine purpose; in other words, form and function are interrelated. We will also bring in the power of a new field of mathematics to understand shapes.

Our focus will be on shapes in nature. It is crucial for both preserving and living in nature to gain a greater understanding of how nature works, and studying the shapes of nature will give us some insight into this topic. The study of nature has also opened numerous doors for civilization, providing the models for advances in science and technology.

Objects can be seen on several different levels. Those on our level in size are the most obvious, such as trees, pinecones, mountains, and insects. But nature also appears at the micro level, for example, in the structure of DNA, and at the macro level, with planets, black holes, and even the shape of the universe itself.

What does all this have to do with mathematics? Some complicated mathematical equations are related to capturing data about pictures, such as mathematical curves drawn on a plane, and for most of us, that's where the relationship between mathematics and shapes ends. But the power of mathematics is that it provides us with a language in which we can understand shapes far, far beyond numbers. As we proceed through these lectures, we will learn new tools to help us understand and manipulate shapes, just as we do with numbers and equations.

The goal of this course is to weave together ideas from nature and mathematics, with special focus on shapes that appear in biology, chemistry, and physics. The course is divided into 4 parts based on the complexity of the shapes. We will first look at **knots**, simple shapes based on lines and circles, which appear in DNA structures, string theory, knotted molecules, and mutations. As we increase complexity, we'll talk about **surfaces**, which appear in space telescopes, stent designs of arteries, and curvatures of mountain ranges. Next, we will move to higher-dimensional objects called **manifolds**. Examples of ideas related to manifolds include soap bubbles and Einstein's theory of relativity. Finally, we will look at superstructures. Here, we will struggle with ideas related to particle motions and collisions, the space of **phylogenetic trees**, and fractals and chaotic systems.

... the power of mathematics is that it provides us with a language in which we can understand shapes far, far beyond numbers.

This course only begins to explore the math underlying these concepts. The first basic mathematical idea we'll look at in understanding the language of shapes is the notion of equivalence. Can we say that 2 shapes, such as a sphere and a cube, are equivalent? In one sense, they are. They both have a smooth, connected surface, but the cube has sharp corners. What about a sphere and a donut? What about a small sphere and a large one? As we'll see in upcoming lectures, equivalence depends on the scale in which we're working and what we care about in our problem. ■

Important Terms

knot: A circle placed in 3 dimensions without self-intersections.

manifold: A generalization of a surface to higher dimensions, where each point on the manifold has a neighborhood having the same dimension.

phylogenetic tree: A mathematical tree structure that shows the relationship between species believed to have a common ancestor.

surface: An object on which every point has a neighborhood that has 2 degrees of freedom.

Questions to Consider

1. In what ways do you see mathematics used in the study of nature, either now or in the past?

2. What shapes are most fascinating to you? Make a list. Do certain common features appear in your list?

Understanding Nature

Lecture 1—Transcript

Welcome and thanks for joining me. My name is Satyan Devadoss and I am a Professor of Mathematics at Williams College. We are about to embark on an amazing journey in the study of shapes. What kind of shapes do we care about? Our ideas can be applied to anything ranging from the kinds of clouds that you see in the sky, whether it be big, puffy clouds or thin wisps, to the types of shapes of mountain ranges, smooth rolling hills, or steep jagged points, to the designs of cars and trucks and airplanes, to even patterns of ties and shirts, like I'm wearing. We want to identify, classify, and quantify these shapes. Why do we care about shapes? The most important lesson that we can learn in these entire series of lectures is that shapes determine purpose. Form and function are interrelated. Consider a coffee cup, a simple object, as an example. Do you know the coffee cup has a little divot on the side where you can put your fingers through so that your fingers don't have to touch the hot cup of coffee? At the same time, the shape of the cup has a groove inside it where you can pour your hot liquid. The form of the shape determines its function.

This doesn't have to apply to something as simple as a coffee cup. It applies to everything you see—to airplanes, to buildings, to trees. Have you ever noticed that there are different types of trees? Some of them are extremely tall and these are usually located in the middle of a dense forest. They're trying to fight their way to the top to get the light in terms of the foliage cover in the forest. On the other hand, some trees are broad, in fields and in plains where they have room and they're trying to expand to collect as much light as they can for photosynthesis.

Shapes are also needed for technology. The most basic digital cameras have facial recognition software. They can look at the shapes that appear in the camera. The software can analyze where a person's face and eyes are so they can focus right to the point, much less compared to the high level of security that's used in airports and by the military nowadays. Mankind has always had a fascination with patterns and shapes appearing in art, design, architecture, and photography. Leonardo da Vinci was a master at this; he blended math, art, and nature seamlessly. You noticed that he never considered at 1 point of his life worrying about art and another worrying about math or engineering. They all blended perfectly together where he never saw the divide between these different worlds.

Near the end of these lectures, we will consider how artists have struggled with the basic ideas of shapes. Most importantly, we are about to bring in the power of mathematics to understand these shapes. The math we work on is completely unlike what is classically considered math and what you might even be comfortable with, which is a good thing. My goal is to take you out of your comfort zone. It's a fairly new field of study, which actually pushes the frontiers of research and hopefully pushes you in new ways. Moreover, we'll look at numerous unsolved problems during our adventures that are absolutely easy to state and understand and has baffled the greatest mathematical minds so far.

This is not meant to intimidate, but to encourage the sense that we can all play the game, each one of us. We don't need to have the training of the great mathematicians and scientists in order to understand the problems. Maybe we'll be able to approach these problems from a fresh perspective, to see things that nobody has seen yet before. This is one of the goals of these lectures. Our focus will be shapes, particularly in nature. This will be our motivation and this will be our guide. You might be wondering, why study shapes in nature when there are so many other things out there? What is it that is so amazing about nature that we should worry about? First of all, it is crucial for both preserving and living in nature to get a greater understanding of how nature works, to actually preserve nature in a time especially like today, and at the same time to understand what nature is really about. To study the shapes of nature will give us a glimpse into this perspective.

The study of nature has also opened us numerous doors for civilization; in particular, there are advances in the sciences and in technology. Let me give you a snapshot of how technology and nature fit together. First, technology tries to understand nature's designs, such as dragonfly wings. It works so beautifully on a dragonfly; why does it do so? What is it about the particular shape of the wing that makes it so efficient on a dragonfly? This is one of our goals, to imitate and understand nature in this way.

Secondly, technology tries to mimic nature's designs. Consider the hypodermic syringe needle. This is exactly motivated and mimicking snake fangs. The designer and inventor of Velcro was motivated by thistle burrs that they happened to walk across in the middle of their forest. These thistle burrs sticking on their jeans motivated this idea of Velcro. Again, we see that technology tries to mimic the nature in front of it. The sonar that we see in submarines is motivated from bats. Even a camera lens that you find in the simplest and cheapest cameras try to mimic the complexity of the human eye.

Nature, it turns out, is ahead of the curve and we are trying to push the frontier of what we can do seeing what nature has already done. The number 1 reason I believe for focusing on nature, as we will see, is that nature is the ultimate optimizer. Let's consider something quite simple like soap bubbles. Have you ever wondered why, when you blow a bubble, it doesn't come in the shape of a cube? Why is it that the bubbles always come, over and over and over again, in the shapes of perfect spheres? Why doesn't the bubble come in the shape of a donut? It turns out that nature is doing its ultimate in optimization. It is using the least amount of surface area, which is the soap film that you're giving to it, to capture the most volume possible. It's capturing the most air. We see that nature is optimizing the best way possible. Thus, we see that nature is an extremely complicated system. It offers us far more than what is obvious at first glance.

I believe that objects can be seen in several different levels. Objects our level in size are the most obvious ones. These are the ones many of us identify with when we say nature—for example, trees, pinecones, mountains, and insects. These are all the examples I've already used today. This is a scale that we can understand and observe and actively influence. Consider something like a mountain or something as small as an anthill. Both of these things are extreme compared to our size, but yet at the same time we can wrap our minds around this. We do understand what an anthill looks like and its size, and we do understand what mountains look like and their size. We can climb mountains, drive around mountains, fly over mountains, bore under mountains; we do have the ability to grasp mountains and anthills.

But, you see nature also appears at the micro level. We see this, for example, in the structure of DNA and molecules and revolutionary ideas like string theory. These things are operating all the time, although beyond our sense of space and distance. We see that in our bodies our cells are replicating over and over again thousands of times and yet we cannot see this operation. DNA is being copied and replicated thousands of times and yet we cannot understand fully this powerful operation at this micro level.

This happens in shapes appearing in the other perspective in terms of the macro level, with planets and suns and black holes and even the shape of the universe itself. We see we can have our perspective where we understand it. We can have the perspective of the shape of nature in a micro level and we can also talk about it in an enormous level. We cannot understand what it means to talk about size and distance at this enormous scale of light years. Yet, nature operates here all the time as well. What does all of this have to do with mathematics? I have been talking about nature for the past few minutes. Doesn't it seem like this kind of a talk or these lectures are designed for a biologist, a chemist, or an engineer? Why a mathematician? It turns out that the problem that we've been worried about has been exposed to mathematics mostly through the language of numbers. Many of us understand that mathematics and numbers go hand in hand. But, it turns out that this is not the case in a big picture setting. It's true, numbers do surround us. The cereal that I was having this morning, on the side of the cereal box, had different percentages of information of my daily iron intake and my daily calcium intake. We see this and we become extremely familiar. I can look at my watch and tell the time of day, tell the date of the year. I can read stock market values and numbers, just pound our brains with these values, weather reports in terms of percentages of clouds or percentages of rain.

Numbers don't just surround us in everything we do, but numbers also define us. Our social security number tells us who we are to the government. Our heights, weight, and birthdays tell us who we are to one another and this is how we hold our information in our minds. We have also become familiar with manipulating numbers, not just surrounding us and not just defining us. We have become masters in manipulating these things, such as simple things like addition and multiplication that we might have learned in elementary school, or even more complicated operations, such as exponentials. We feel good about numbers and equate this with what mathematics really is about. Mathematicians must be masters at manipulations of numbers.

When we think of more complicated or advanced mathematics and we pause, we think about the world of equations. Maybe the great mathematical results are the ones involving the Pythagorean theorem, $A^2 + B^2 = C^2$, or the quadratic formula, $-B +/- \sqrt{B^2} - 4AC/2A$. Wow, can things get more complicated than this? What about trigonometry or maybe even calculus equations? Derivatives and integrals come to mind. What does this have to do with shapes? Some of the equations that we have mentioned might be related to capturing data about pictures. Maybe an equation such as $X^2 + Y^2 = 4$ relates to capturing what the picture of a perfect circle is of radius 2. Maybe this is how mathematics and shapes are related. Maybe it's related to equations of parabola and other more complicated mathematical curves that we can draw on the plane.

But, to most of us, this is where many of us end in our pursuit of how mathematics fit with shapes. But, the power of mathematics is that it provides us with a language in which we can understand shapes far, far beyond numbers. Consider something as simple as a coffee cup. How would you describe something like this? What kind of numbers would we use to understand this kind of a shape? Or, how do we describe a cloud, the big puffy ones or the thin wispy ones? Would you talk about the length of the cloud or the volume that the cloud takes up? That somehow doesn't get to the heart of the shape of the cloud itself. What about a tree? Maybe we can talk about the branches of the tree or count the leaves or talk about how the tree approximately looks like maybe a parabola or maybe certain trees look like circles. But, again, we're losing the essence of what the tree is really about. It turns out that numbers and equations are not enough. We are not fluent in this language of shapes. We're trying to say something and yet we haven't learned the words to speak this language.

A natural thing to say is, why not talk about a biologist or a chemist or a physicist? Who should talk about these lectures? It turns out that mathematicians are the ones unlike these other scientists. We are the ones who have created and who speak this language fluently. Something like a tree, which we see all the time, becomes hard to describe without this language. The whole of the tree turns out to be far greater than just talking about the leaves and the roots and the bark itself. In this course, we will learn to speak this language over these lectures that are following. We will also, at the same time, learn tools to understand shape and to manipulate the shape, just like we're able to do with numbers and equations. Remember how we can take 2 numbers, 3 and 4, and put them together and create a new number, 7, using addition? We can take 3 and 4 and put them together and get a new number, 12, using multiplication. Can we do this with shapes? Can we take 2 different shapes, combine them in a certain way, and get a new shape? Are there different ways of doing it? What are the consequences? What are the uses? These operations that we learned in numbers, can they be extended to shapes?

In this course, we will use any weapon we can fashion to break through our obstacles in understanding shapes—whether it's a number, we'll use it, whether it's polynomials that we learned in algebra, we'll use it. What about coloring from kindergarten? We'll use it. There is no idea that is too big or too small for us to use for our understanding of shapes to improve using the language of mathematics. There is more to math than providing us with a language. As we learn this world of mathematics, we find that we're only at the tip of true understanding. Most of math remains a mystery even today. If this is the bar of understanding of math since the beginning of time to where we are at the end of time, where would you say most people would rate our understanding of math? I believe most of the people you would talk to would say that we are here. We're nearly fully able to understand all of mathematics that there is. But, if you ask a mathematician, if this is the math at the very beginning and all the math knowledge there is, they would say we are right here at the very beginning of understanding.

We are basically kids in a playground, in a sandbox, learning how to use a rake and a shovel. The tools of Pythagoras and the tools of the quadratic formula are at the tip of the iceberg to all the math that's out there that's yet to be discovered. Each of the results that we have opens up 7 more doors than it closes. As we solve the math problem, as we're able to dive in, we see that things are far deeper than they appear. Most of the math that we're exposed to is from an ancient era, from the Greeks, the Egyptians, the

Chinese, hundreds and thousands of years ago. The progress of math is slow based on building on previous works.

All of us have heard about the world of geometry, the math having to do with shapes. A new branch of math, relatively from the perspective of time, is called topology, which basically gets to the heart of shapes. It's only in the 1800s and the 1900s that this entire new branch was really studied in detail. Even today, numerous unsolved problems lay dormant in this world, which we will learn and talk about. At the same time, we'll learn about numerous unsolved problems, not just in this world of topology, but across the way math and nature fits together. These, thus far, have been impossible to prove by mathematicians or by any other scientists working in this world.

What makes a great mathematician? The great ones can solve or make immense contributions to practically any problem they choose. However, they will be judged on not just how they solve the problem, but the area of research that they have decided to focus on. In particular, the great ones are opening doors and breaking down walls and determining the direction of math research for the next 50 years. This is what makes them great. Moreover, the great mathematicians have not only excelled in understanding what is given in front of them, but to make connections between what was thought to be practically unrelated fields. In math, the greatest honor you can get is not the Nobel Prize. There is a rumor that a mathematician had an affair with Alfred Nobel and things didn't work out. Thus, the great honor of mathematics is something called the Fields Medal. This is in honor of John Fields. It's given every 4 years and 1 must be not over 40 years old in order to get it, unlike Nobel Prizes, which you can get in your 60s, 70s, and 80s after your ideas have been established for many years. To get a Fields Medal, you must have revolutionized the field when you're still a kid.

We will look throughout these lectures at several Field's medalists. What makes a great mathematician and what makes great mathematics? It is the power of abstraction. Let me give you a simple example. Mathematicians in the early 1900s began to study the symmetry of objects, such as a cube. They would take a cube and, if you rotate it, you get the same object again. If you rotate it this way, you get the same cube again. This study of symmetry was abstracted by mathematics and by mathematicians to this world called group theory, which we will study in these lectures.

Group theory is the study of symmetries and structure of mathematical ideas, not concrete objects. Many non-mathematicians thought that this was absolutely silly, not having a direct application to the real world. However, group theory was used decades later for quantum mechanics. Things which are abstract will be useful for things we never thought possible. Moreover, it might not give us a better car, or a better toaster today, but it will lay a foundation for a better toaster someday. Most importantly, mathematics is pursued for its own sake. It needs no defense. Of course, this is coming from my absolute unbiased perspective.

The goal of the course is a 2-stranded weaving of ideas of nature and mathematics. Sometimes we'll venture more into the math side than the nature side, mostly since I'm a mathematician at heart. The key interest will be in shapes that are appearing in biology, chemistry, and physics that are motivating our work. For example, in biology, the shape of leaves will be related to origami designs, how leaves sprout out based on certain folding properties, determining when to intake light. In chemistry, the shape of molecules determines functions of polymers, enzymes, and the DNA itself. In physics, the shape of space and even time will determine how light and gravity work together.

The structure of this course is divided into 4 groups based on the complexity of the shapes that appear. The simplest shapes are based on lines and circles. Our main concern here will be the study of knots, how we can take a simple piece of string, tie a knot in it, and these appear in DNA structures, string theory, knotted molecules, and mutations. The second step of increasing complexity we'll be talking about is surfaces. Everything your eyes see is actually capturing a surface of an object. If you're looking at me, you don't see all of me, you don't see my heart or my lungs, you don't even see most of me at all. You only see the surface of the clothes that I'm wearing. If we understand surfaces, it'll go a great way on how we see and understand the world around us. These appear in origami designs, of space telescopes, stent designs of arteries, curvatures of mountain ranges, tilings of honeycombs, and animal patterns. One of the examples we could even do is to talk about rigidity. For example, taking simple rods of balls and joints can be imitated by protein modeling. We can talk about wind flow patterns, how the wind flows on the surface of the earth and how the shape of the earth itself is determining the cyclones and the types of tornados that appear. After we understand simple shapes like knots and move on to surfaces, we can talk about higher dimensional objects. These are called manifolds and are quite difficult to understand even today. These are hard to understand because, unlike surfaces that we see, manifolds are bigger than us because we are in them. They surround us. Examples of ideas of manifolds are related to soap bubbles, packings of clusters, and Einstein's Theory of Relativity. We can even be ambitious enough, which we will be, to talk about the shape of the universe itself.

We are going to close these lectures with shapes of super structures. These we will try to capture in a metaworld, a shape of shapes. We will struggle with ideas related to particle motions, particle collisions, the space of phylogenetic trees, fractals, and chaotic systems, and how these are related to weather forecasting—not just in nature, but economic forecasting conditions as well. Near the end, we are going to talk about visualizing and drawing and understanding higher dimensions.

Let's take a look at some of these examples here. This is a snapshot of a 4-dimensional object that's been pushed into 3-dimensional space. What does that even mean? How can we make this or understand its properties? Let's take a look at this example. This is a Calabi-Yau manifold, a 6-dimensional object that is again being pushed into 3-dimensional space. What is the purpose of this and what is the point of us understanding something so complicated? Calabi-Yau manifolds appear in string theory, which tell us the foundations of the world itself. This course only begins to explore the math underlying these concepts. However, we will learn basic tools and understand how mathematicians think when they attack these problems. The first idea in understanding the language of shapes is the notion of equivalence. We're going to explore this notion more in detail in the next lecture, but let's jump in to get a taste of what this is about. What do we mean when we say that 2 shapes are equivalent? For example, what does it mean for me to say 2 vehicles are equivalent? Does that mean that these 2 modes of transportation must be of the same kind? Maybe they need to have the same number of wheels. Maybe to say 2 things are equivalent means they must have the same engine or maybe they must be the same make and model or even the same color.

How can we tell things apart if we have no means of measuring equivalence? Who determines what is or what is not equivalent? This is the beauty of mathematics. We determine what's equivalent or not. Mathematicians define different levels of equivalence. We determine whether a ball and a cube are equivalent. Let's think about this. If I have a cube in one hand and a sphere, a perfect ball, in the other one, are they equivalent? In one sense, yes they are. They both have this smooth connected surface on one side and practically a relatively smooth connected surface on the other ones; they both have no punctures in them. But, one has these sharp corners, so you feel that it probably isn't equivalent to the other one. We again determine whether they're equivalent or not. We can talk about whether a ball is equivalent to a donut. In both cases, you have smoothness unlike the cube and the sphere, but here the donut has this intrinsic hole in it that the ball doesn't. Maybe they should not be equivalent objects. Something makes us say no, but it's not just the connectedness. It's something deeper and here we see us speaking without this mathematical language. What about a big ball and a small ball? Are these equivalent? Again, we need to know more than just wondering about volume or surface area. We need to get to the heart of what equivalence means.

Equivalence depends on what we care about. It is measured by the lens through which we view our problem. Different areas of nature call for different lenses. In situations of micro, macro, and normal scales, we will use different notions. Each kind of equivalence has its strengths and weaknesses. There is no one answer that always works; it depends. Over the next few lectures, we're going to understand what equivalence means and the different kinds of equivalences we're going to worry about. What have we talked about today? We have begun our adventure on the shapes of nature. We've talked about how form and function in nature are intrinsically related, and understanding the shape of it will help us understand the function of the object we're studying. We've also seen that nature comes at 3 levels, not just what we can experience, but in an extremely small micro level to an extremely large macro level. Mathematics provides a language to understand these complicated structures. Thank you and stay tuned.

The Language of Shapes
Lecture 2

Topology is a part of geometry that focuses on the qualitative—not quantity but quality—in other words, the kind of position and the kind of underlying structure that defines the object.

We can think of mathematics as broken into 3 classes: analysis, algebra, and geometry. Analysis is the area of mathematics that studies change, for example, the growth of a given population. Algebra is used to study structure, and geometry is the study of shape. Of course, mathematics is far more complicated than this breakdown would imply. There are also combinations and specializations of these fields.

Topology is a relatively new branch of geometry that focuses on the qualitative, in other words, the position and underlying structure that define an object. In particular, the word to associate with topology is "relationships." The origins of topology can be traced back to the work of **Leonard Euler** and the problem known as the Seven Bridges of Königsberg. Euler's work of genius was to convert the map of the city of Königsberg into a graph. He showed that it was impossible to connect every region of the city with an even number of edges—one going in and one coming out. In solving the problem, Euler ignored concepts of area, length, volume, and angle, but he talked about relative position, how things were connected.

To create a finer distinction between geometry and topology, consider equivalences of shapes. Equivalence in geometry is based on rigid motions. For example, if we rotate an object, it is still the same object as it was originally. Equivalence in topology is based on a world called isotopy, sometimes called rubber sheet geometry. Here, we're allowed to stretch and pull an object, but we're not allowed to cut and paste. If we take an object and stretch it like clay, relative position does not change. Thus, from a topological point of view, a cube and a sphere are equivalent.

Different situations call for study through geometry or topology. Macro and micro problems are usually too hard for geometry but are possible for topology. We cannot answer questions about size, volume, and area in a macro or micro setting, but we can grasp relative position.

In closing, we look at the concept of **dimension**, which is simply a number that we associate to a shape. Dimension is broadly defined as degree of freedom—the smaller the dimension, the less freedom you have, and the greater the dimension, the more freedom you have. Dimension is narrowly defined as the amount of information needed to pinpoint a location. An example of a zero-dimensional world is a point; a line and a circle are both 1-dimensional, the plane is 2-dimensional, and our universe is 3-dimensional.

Equivalence in topology is based on a world called isotopy. This is sometimes called rubber sheet geometry. Here, we are allowed to stretch and pull on our object, but we're not allowed to cut and glue.

The important point to remember here is that dimension is simply a mathematical construct; it associates a number to a shape. Thus, dimension doesn't have to stop at 3. Maybe our imaginations stop at understanding things beyond 3 dimensions, but dimension does not need to stop there. In the next lecture, we'll begin a further exploration of dimension with 1-dimensional objects called knots. ■

Name to Know

Euler, Leonhard (1707–1783): One of the greatest mathematicians of all time, his scientific works cover analysis, number theory, geometry, and physics. He was one of the first to use topology, from which we receive the formula $v - e + f = 2$ of a polyhedron.

Important Term

dimension: An invariant given to a point on a shape that measures the degrees of freedom afforded at that point.

Questions to Consider

1. Which feature of an object, its geometry or its topology, do you seem to notice instinctively? Why?

2. What is the dimension you are most comfortable dealing with? Is it three dimensions, given that you are a three-dimensional being? Or is it one or two dimensions?

The Language of Shapes

Lecture 2—Transcript

Welcome back and thanks for joining me again. Let's start by reviewing what we did last time. In the last lecture, we talked about why we are interested in the study of shapes, especially in the study of shapes as it pertains to nature. We noted that this is useful for everything we do, from arts to technology. Whether we are designing the next high speed airplane or the 1 that relates to space shuttles, flying in space, art, nature, and mathematics are all bound together. In particular, the most important thing we learned was that form and function are related. In other words, the more we start understanding about an object's shape, the more we start understanding how this works together. Third, we realized that nature was an amazing optimizer. We saw this when we related the shape of bubbles, simple soap bubbles. We saw that nature tried to optimize the most volume it can based on the given amount of surface area, the soap film that you have.

Finally, we saw that nature is more than what we see. It is very easy for us to say that nature is about trees, plants, and mountains. But, nature operates at 2 other levels far beyond our understanding. One is at the micro level, at the level of DNA structures, cell designs, atoms, and molecules. It is even related to ideas we're going to talk about motivated by strength theory. At the same time, nature also relates at the macro level, in worlds that are far beyond us in terms of galaxies and star clusters. Here, the concept of distance isn't extremely small, it's extremely large. We're talking about light years away to talk about distances and movement.

Moreover, we discussed the power of mathematics in terms of how nature fits together in understanding its shape. We noted that mathematics provides us with a collection of tools. Mathematics is more than just numbers; it's actually a language, a way of communicating and describing. When we try to understand how to describe a shape as simple as a tree, we see that our language isn't enough. We come to a barrier and mathematics helps us to break through this barrier, to understand what shapes are about. This course is broken down by dimension, an idea we're going to talk about later in this particular lecture. We will go through increasing levels of complexity as we progress throughout this course. Finally, what we talked about last time was a concept of equivalence. We framed this with the idea of talking about geometry, the world of mathematical shapes and this new world called topology that broke through mathematics around 1800 and 1900. We're going to jump into more detail this time in terms of understanding topology and what it means to talk about equivalence. Mathematics itself is broken into several main fields, each with its own set of tools. What I'm about to give you, I'm going to admit, is my personal perspective on the way math is broken down. I must say, though, that almost any graduate school you attend today in mathematics will force you to take 1 of 3 classes in the areas of analysis, algebra, and geometry. Therefore, that's the breakdown that I'm also going to use to model what it means to break mathematics into pieces.

Let's start at the very first one. What is analysis? Analysis is the area of mathematics that studies change. For example, if you're interested in understanding how the speed of an object has changed—whether it's the speed of a leopard or the speed of a bullet—the way to measure and quantify this change is through the lens, the world, and the language of analysis. Maybe you're interested in the rate of growth of population, the growth of humans over time, or the growth of bacteria in a Petri dish. Again, we turn to analysis to measure this growth, this rate of change. Maybe what's interesting, especially nowadays, is to understand the stock market as the world is built on these financial models. To understand stock market fluctuations, whether up or down, to measure these chaotic changes, again uses the tool of analysis. This is the world in which we see. Measuring global warming, not just over the span of a few hours or days, as you would do for a stock market, but over millions of years, again uses analysis. Time is not an impediment to understand how analytic systems work for a mathematician. That's the first main branch of mathematics.

The second one, the second superstar way of breaking mathematics into 3 pieces, is algebra. Those who are superstars in algebra, those are the great mathematicians who understand how to study structure. Whereas analysts talked about change, an algebraist is interested in the details of the operation, a meta way of thinking about it. Let me give you an example. For an algebraist, operations on numbers are what are exciting, such as addition and multiplication, not the numbers themselves.

An algebraist looks at the real number line, pulls back, and sees the bigger picture between numbers. What are prime numbers, the way you put numbers together, 7×1, that there's only 1 way to do that? The fact that there's a structure of what addition and multiplication does to numbers is what makes an algebraist tick. We can go behind the scenes to get a bigger picture in terms of how algebraists work. Remember the example we talked about last time—taking a cube and rotating it? What's interesting to an algebraist isn't the cube itself, but the ways the rotations, 1 way or another way, can occur. The structure of the cube is what algebra is about.

The third way of decomposing math into its main frameworks is geometry. We've all heard of this. Geometry is the area that studies shape. This could be very classical in nature, what everyone is comfortable talking about. Whether this is length, distance, area, volume, or size, all of these are geometric ideas. We can even use ideas in trigonometry—sine, cosine, and angles and we're still in the world of geometry. On the other hand, this could be cutting edge ideas in geometry that are in a revolutionary new setting, which we will talk about later.

It's very superficial for me to say that there are only 3 worlds of mathematics—analysis, algebra, and geometry. Nowadays, it's far more complicated than this. Note that there are also combinations and specializations of these fields. Let's take a world called number theory. What is number theory? Number theory is a branch of mathematics that's interested in how numbers work. This is a subfield of math, which is in the world of algebra and in the world of analysis. A number theorist can use the lens of algebra to understand how number systems work, to see the structures of numbers. They can also use the power and the tools of analysis to see how the numbers change as you go through the number system. We will use tools from analysis and algebra for our work in geometry. As we talked about last time, we'll use any tools we can get. Geometry is a language which focuses on the quantitative, on the amount in which we measure shapes. Notions such as the length, area, and volume deal with these geometric ideas. The study of geometry has been around since antiquity and this is one of the main reasons we teach geometry in all the school systems.

The Egyptians, in 2500 B.C., came up with an accurate approximation of the area of a circle with less than 1% error, a stunning result in a geometric setting. The Babylonians, in 2000 B.C., created general rules for measuring volumes of solids. The Indians, in 800 B.C., came up with geometric approximations of the $\sqrt{2}$. The Chinese, in 300 B.C., measured areas and volumes of numerous objects in very special ways. The Greeks, at the same time, in 300 B.C., came up with ideas motivated and promoted by Euclid. In Euclidian geometry, all geometric structure is now viewed through the point of construction, starting at points, moving to lines, and based on a collection of axioms. In 700 A.D., the Middle East nations created a mixture of algebra and geometry—again, to provide powerful new results never seen before.

All our lives, the focus for studying shapes has been on geometry throughout the schools. We learned about areas of triangles and circumferences of circles. Even all of calculus focuses on the geometry—the shape of curves and surfaces where length, area, and volume matter. Of course, I don't want to shortchange geometry because it is fundamental and foundational for calculation and precision.

You need geometers as engineers where we actually need to know the angle, length, and area of the bridges we're building. It is also the key to our size space. When we talk about mountains, when we talk about anthills, geometry is the kind of language we need to compare 1 to another. But, geometry is quite difficult to use in the micro and macro worlds. We'll talk about this in a little bit. What about this world called topology, this new branch of geometry that started in the 1800s and the 1900s? Topology is a part of geometry that focuses on the qualitative, not quantity, but quality; in other words, the kind of position and the kind of underlying structure that defines the object. In particular, the word to associate with topology is the word "relationships." Unlike geometry, this is a fairly new discipline. It can be traced back to the work of Leonard Euler in 1736. Topology and the way 1 would think topologically has existed before. It's a very intuitive thing, but it was actually foundationally written and published in Euler's work on the Seven Bridges of Königsberg.

Instead of just telling you what topology is about, let me actually show you by digging a little deeper into Euler's work. Here's the city that Euler cared about, Königsberg in Prussia, which is now in Russia. This city was set on both sides of the Pregel River. It included 2 large islands, which were connected to each other by 7 bridges. Take a look at this picture. Here, we see the central island in the middle, an island on the side, and 2 mainland masses on the top and the bottom. Notice the connections of these islands and land masses by these 7 bridges.

This is what Euler had to work with. Here is the problem that was given to Euler and to the residents in Königsberg. Imagine a friend of yours is a tourist and he or she is coming to visit you. What they want to do is to explore this city. But, if you look at the picture of the city again, notice that what's so beautiful about this city are the bridges, these 7 beautiful bridges linking up different parts of the city. What you want to do is to take your friend on a walk, such that you go through the city where each bridge is crossed on your walk once and exactly once. You don't want to cross the same bridge twice; that would bore your friend. On the other hand, you don't want to miss any of the great attractions; you want to make sure every bridge is hit.

How do we do this? First of all, notice that we can start anywhere in the city, it doesn't matter what point we start at. If you think about it, we see that it's not based on analysis or algebra. It's based on geometry; Euler's really interested in shape. He's not interested in a deeper structure. We see the structure obviously in front of us. He's not interested in change of movement. It's based on this picture itself. Euler's work of genius was that he converted the map that you see here to another map, which we call a graph. It lost all concept of geometry in terms of distance, but it kept its relative position, and that's how we enter into topology. Let's take a look at Euler's version. Notice what Euler did. At every region of the city that is separate and disconnected, Euler uses a little dot, which we call the vertex. That is what's important about that region. But, what he emphasized was the bridges. He deemphasized the regions because walking around within a region is not exciting to this particular problem. He emphasized the bridges because that's the key to the problem. His previous map of Königsberg has now been replaced with this other map of Königsberg.

He proved that there was no solution to this problem. It's impossible for you to take your friend on a tour of the city that hits every bridge exactly once and make sure you don't cover any bridge twice to bore your friend. Let's see how Euler proved this. He said, let's start at any vertex we want, at any region of the city. Here's the way his thinking worked: if you're not starting or stopping at any other region of the city, other than your start and end positions, then you must have an edge that goes into that region. At the same time, you must have an edge that leaves that region. If you're not starting and stopping at this edge you need to get in and you need to get out.

What does this mean? It means that all the vertices, all of these dots which represent these regions of the city, must have an even number of edges coming in and going out—except, possibly, at 2 start and end positions. Except for these 2 start and end positions, the number of sticks coming out of a vertex has to be even. Notice that every particular vertex is odd. Number 3 sticks on that one, 3 on that one, 3 on that one, and 5 in this one. We see that it's impossible to do so. No matter what you pick to be your starting and ending position, it's impossible to go on this walk. Euler ignored concepts of area; we never mentioned it in our proof of this problem. He ignored length, volume, and angle, but he talked about the relative position, how things were connected.

You see that the problem is still about shape. It's still about shape, but in a topological setting, not in a geometric setting. This is the world of topology where ideas of connectedness and continuity become the key. This field was not fully recognized until the end of the 19th century—many, many years after Euler pushed this result forward. At that time, it was called "geometria situs" in Latin, which means geometry of the place, or geometry of position. To distinguish between these 2 areas even more, consider equivalences of shapes in geometry and topology. Equivalence in geometry, what we talked about last time—in terms of what it means for objects to be equivalent—is based on rigid motions. For example, if somebody gives me an object, I can rotate that object and still consider it to be the same object as before. I can say that those 2 objects are equivalent because they're just off a little bit by rotation.

Maybe I can take an object and reflect it and say that's equivalent. For example, my right hand and my left hand would be considered equivalent because they are mirror images of one another—they're not too far off. Maybe translation is equivalent to our eyes, where you take a cube here and you can translate it to this position and still call it an equivalent cube. These are all notions of congruence where 2 elements are equal based on rigid motion. Maybe at the worst extreme level of equivalence in the geometric perspective, we can talk about change of scale. Maybe a small cube can be scaled into a big cube and we can see both of those are equivalent.

Notice that volume is definitely lost, but the geometric ratios of the objects are preserved. To a geometer, a right angle triangle—which has a 90 degree corner and 2 45 degree corners—and an equilateral triangle—which has 3 60 degree corners—are very, very different. This is good because we need geometers to build bridges and to build the world in our system. But, it could actually get in the way, like in the Königsberg bridge problem. If we start thinking about area, volume, and distance, we see that this actually impedes us from really understanding the heart of the matter.

Equivalence in topology is based on a world called isotopy. This is sometimes called rubber sheet geometry. Here, we are allowed to stretch and pull on our object, but we're not allowed to cut and glue. If I take an object and stretch it like clay, the relative position has not changed. Of course, things are getting further away than before. But, the things that used to be further away, originally, are even further away. Things are relatively the same. Things that are close relatively stay close and things that are far relatively stay far. Ideas of connectedness and continuity are important. Let's consider some examples. From a topological point of view, is a cube and sphere topologically equivalent? Let's think about it. If somebody gives me a cube, I can take the cube and, with isotopy, or rubber sheet geometry, push the corners of the cube in and make it into a sphere. In this case, I can make the cube into a sphere because the concept of relative position is maintained. Let's try to classify some of the letters of the alphabet under this notion of isotope. Here we have the letter C. If I take the letter C, what is it equivalent to from a topological perspective? One thing I can do is take the letter C, straighten it out, and make it into the letter I. I haven't done anything other than stretching. One thing we cannot do in topology is cut and reglue. By cutting, I'm making things that are close, now very far. By regluing, I'm making things that are very far, close. I've changed relative position.

Notice that C is equivalent to I, which is equivalent to J, which is equivalent to S, and numerous other letters. What about the letter F? We consider here and look at the letter F; we see that the letter F cannot be made into the letter C. Why not? The letter F can be made into E quite easily. It can also be made into T or Y quite easily because I can stretch and pull. But, I have to eliminate this part where there are 3 branches coming out of this position. That captures some data for me that has this concept of relative position of these 3 different places coming in together.

Think about the Königsberg bridge problem again. You need that vertex where 3 possible things meet. This idea of equivalence between geometry and topology does not mean that 1 is better than the other. For example, if we want to compare the letter C to I and J, that's great. If we want to compare F to E, T, and Y, that's great. But, the letter A is distinctive too. Note that in the letter A, we have the central circle and these 2 branches. We see that A is equivalent to D, which is also equivalent to *r*, or to these other crazy shapes we can create just by stretching and pulling. In particular, different situations call for study through geometry and topology in different ways. Macro and micro problems are usually too hard for geometry, but it's actually possible for topology. Why is this? We cannot answer questions about size, volume, and area in a macro or micro setting because this is too difficult in many cases. But, we can grasp relative position. Since topology asks for a weaker condition—like the Königsberg bridge problem didn't ask for a more difficult problem of geometry, but a weaker notion of topology—it's the right tool for us to do it. Let us close these lectures by turning to a concept of key importance called dimension. Dimension is the overarching characteristic by which we're going to break down the structure of this course. We talked about 1-dimensional objects, like lines and circles, 2-dimensional objects like surfaces, and 3-dimensional objects like the shape of the universe.

What is dimension? It's a buzzword we hear all the time. We talk about 3-D glasses to see the new movie in. We talk about time being the mysterious fourth dimension. What is it? Dimension is simply a number that we associate to a shape. We broadly define this as the amount of freedom you have—the smaller the dimension, the less freedom you have and the greater the dimension, the more freedom you have. We narrowly define this as the amount of information needed to pinpoint your location. Imagine you're in a 0-dimensional world. What does this mean? What would this look like? It means that you know where you are and everybody knows where you are without giving any information. Zero pieces of information are needed to know where you are. But, the only place no information is needed is if you lived at a point, at 1 spot. Thus, you have no amount of freedom to go anywhere. This, a simple vertex or a point, is a 0-dimensional world.

What about 1 dimension? You need a 1-dimensional piece of information, which means you need a piece of information for every point that you are. But, only 1 piece of information is needed. Consider the real number line. This is an example of a 1-dimensional world. Say you live in house number 73. Your friend calls and asks where you are. You say, I'm at 73. They know exactly where you are with that 1 piece of information. Your friend lives at house −4. You can find out exactly where your friend is with that 1 piece of information, −4.

What about a circle? A circle looks very much like a line, but it looks like it's also a 2-dimensional world. Although it takes 2 dimensions to draw the circle, because it has width and height, it turns out the circle is still a 1-dimensional object. Why? This is because, if you lived on the circle, you only need 1 piece of information to tell somebody where you lived, just the angle in which you live. You can say, I live on 14 degrees. I live on 178 degrees and that tells your friend exactly where you are on the circle.

What about 2 dimensions? An example of the 2 dimensions is the plane. We were very familiar with this when we were in school and we talked about the X-coordinate and the Y-coordinate, the XY axis. This is 2 dimensions because you need X, 1 entire axis, to keep track of where you are in this direction. You also need Y, an entire axis, to keep track of in this direction. Any point on the plane needs 2 pieces of information, your X-coordinate and your Y-coordinate. What about a sphere? A sphere, again, looks like it's a 3-dimensional world because it does have height, width, and breadth. Yet, a sphere is only 2 dimensions. If you live on the surface of the sphere, you see that you only need 2 pieces of information, longitude and latitude. That pinpoints your location on the sphere.

What about 3 dimensions? Consider our universe. Our 3-dimensional universe needed 3 pieces of data. You need to know your X-coordinate, your Y-coordinate, and your Z-coordinate—width, height, and depth. But, what I want to emphasize more than anything is that dimension is simply a mathematical construct. It associates a number to a shape. You see a shape. You can give a number to it. You see a sphere and you can give the number 2. You see a plane and you can give the number 2. You see a circle, you give the number 1. Thus, dimension doesn't have to stop at 3. Maybe our imagination stops at understanding things beyond 3 dimensions, which I will try my best to fix. But, dimension does not need to stop at 3. You could imagine a 17-dimensional world where you need 17 pieces of data to pinpoint your location. Near the end of these lectures, we will push into a world that will need hundreds and hundreds of dimensions. It's a beautiful thing.

In conclusion, what have we talked about today? We have continued our adventure on shapes. We started with looking at mathematics in a big picture setting—seeing analysis, the study of motion and change, algebra, the study of structure, and geometry, the study of shapes itself. We actually zoomed in a little bit in geometry. We saw that, within geometry—which cared about shape, distance, area, and volume—there's a subworld that studies shapes called topology. It's a brand new, young world and there are numerous problems that we're going to dive into.

We also learned that topology is probably the most useful tool in talking about micro and macro worlds—unlike geometry, which is fantastically useful for the world we live in. We noted that geometry has a feel of quantitative study while topology has a feel of qualitative study, something that has to do with relative position. We also considered equivalence of shapes in topology, where we can talk about the letters of the alphabet or a sphere becoming a cube. We closed with talking about dimension itself, what this great buzzword was and its simple idea of associating a number to a shape. In our next lecture, we begin our adventure with 1-dimensional objects called knots. Stay tuned.

Knots and Strings

Lecture 3

We've proved a truly beautiful result based on shapes. It's based on Reidemeister moves, which all are local phenomena that control a global structure. It is based on colorings—simple ideas that we had when we were kids, applied in a powerful way. This is the type of creativity and originality that pushes the frontiers of math.

In this lecture, we consider the simplest and most elegant of shapes, 1-dimensional ideas of circles. For a topologist, a circle is simply of piece of string in which the ends are connected. Our goal is to study how such circles sit in 3-dimensional space.

The most beautiful examples of circles in 3 dimensions are knots, which are defined as circles that are placed in 3 dimensions in different ways. The most elegant way to construct knots is simply to close up the ends of strings. The simplest form of knot is a circle, called the unknot. Another knot is the trefoil. By isotope—by just stretching and pulling without cutting—we cannot make a trefoil into an unknot. Thus, these 2 objects seem to be very different topologically.

The best way to study knots is to look at their projections on paper. The tools needed to study projections were provided by Kurt Reidemeister, who developed 3 moves that change the knot projection but not the knot it represents. According to Reidemeister move I, we can twist a vertical string one way or the other and still have the same knot but with different projections. With Reidemeister move II, we can push one vertical line behind another, introducing 2 new crossings and, again, changing the knot projection but not the underlying knot. With Reidemeister move III, we can move any strand that's behind or in front of a crossing below that crossing. In other words, a crossing at the strand above or below is independent of where it's placed. These 3 moves are the only ones needed to go from one projection of a knot to any other projection of the same knot, regardless of how complicated we make the sequence of the 3 moves. Reidemeister moves can help us tell whether 2 knots are the same, but they do not help us tell knots apart.

A major breakthrough using Reidemeister moves was introduced by Ralph Fox in the 1950s with the idea of coloring. According to Fox, a projection of a knot is 3-colorable if it meets 3 conditions: If we can color every strand (a piece of knot that goes from one crossing to another) using one of 3 colors, if all 3 colors are used in the knot projection, and if either all 3 colors or only one color meet at each crossing. We see some projections in which the unknot is not 3-colorable, but the trefoil is.

The most beautiful examples of circles in 3 dimensions are what we call knots. Knots are simply defined to be circles that are placed in 3 dimensions in different ways. The most elegant way of constructing knots [is] by taking strings and just closing up the ends of the string.

How do colorability and Reidemeister moves fit together? According to Reidemeister, if one projection of the knot is 3-colorable, then every time we do a Reidemeister move, it stays 3-colorable. Further, every possible projection of this knot will be 3-colorable. In other words, 3-colorability is not a property of the projection of the knot but a property of the knot itself. The idea of 3-colorability allows us to see that, in fact, the unknot and the trefoil are different knots. ■

Suggested Reading

Adams, *The Knot Book*.

Questions to Consider

1. How many different knots can you draw on paper where the knot crosses itself five times or fewer? What about crossing six times?

2. Can you generalize the coloring method to use four colors rather than just three?

Knots and Strings
Lecture 3—Transcript

Welcome back and thanks for joining us today. We have continued our adventure on shapes for the past few lectures. Let's see what we had thought about. Do you remember how we struggled with thinking about the different branches of mathematics—analysis, which has to do with how things change, algebra, which has to do with the structure of mathematics, and geometry, which has to do with shapes? We got a glimpse of how these worlds worked, kind of in a big picture setting. As we progress through these lectures, we'll dive into each one of them again, from the perspective of wanting to study shapes more. We also understood that there's this relatively new branch of mathematics that started around the 18–1900s. This was called topology. The focus of topology was still studying shapes, but it was not about area, volume, and distance; it focused on relative position. Thus, our concept of what it meant to be equivalent from a topological perspective is different than what it means to be equivalent from a geometric perspective. This new idea of equivalence is useful when we talk about things at different levels of the world—at the micro level when we look at DNA structures, or even structures of atoms, even smaller to super string theoretic ideas, as we'll discuss later, or to the macro worlds when we talk about black holes and galaxies. Both of these extreme cases are well-designed for us to use our new tool of topology to understand it.

We also talked about equivalence of shapes in topology and we called this isotope or rubber sheet geometry. Do you remember how we took the letters of the alphabet and we can stretch it and pull it and change it and have a new concept of equivalence based on topology? We closed our ideas with the study of dimension, when we talked about dimension as a number that you associate to a shape to measure how much amount of information you need to tell somebody where you live on that shape. Today we begin our jump into shapes by considering the simplest and the most elegant of these shapes, 1-dimensional ideas of circles. Remember last time—all those circles are 1 dimension and it might take 2 dimensions for you to draw. It even might exist in space in a 3-dimensional perspective. What is a circle for a topologist? It's simply a piece of string that you connect the ends up together. This is all circles are. Our goal is to study how these circles sit in 3-dimensional space.

Remember, whether I take the circle and I put it on the plane, which is 2-dimensional, or I take the circle and I hold it in my hand and move it around in 3 dimensions, where it does have X, Y, Z coordinates to draw it, notice the circle is still a 1-dimensional object. This is because if you lived on the circle, there's only a 1-dimensional world of possibilities in which where you could be. Anywhere you are, you could tell your friend where you live with that 1-dimensional piece of data—walk 13 feet to the left to find me or walk 4 feet to the right to find me. One piece of information is enough even if you're in 3 dimensions.

What are the properties that we can do with circles in 3 dimensions? The most beautiful examples of circles in 3 dimensions are what we call knots. Knots are simply defined to be circles that are placed in 3 dimensions in different ways. The most elegant way of constructing knots are by taking strings and just closing up the ends of the string. Consider some examples. Here is a piece of string. As I close it up, it forms a circle. This is the simplest form of knot possible. Since it's instinctively not even knotted up at all, we call this the unknot.

What else can we do? We can take the piece of string and tie a shoestring knot, what you'd use to tie your shoelaces with—probably the simplest knot that you can see like this. Now close the ends of the string together. This thing is again a circle that we have because it starts and ends at the same place. However, it's been placed into space in a different way than the unknot has. This piece of knot is called the trefoil. Let's take a look to see why. If I place this on this flat plane, we see that there are exactly 3 bumps that are based on this knot. Since there are these 3 bumps, we get the word tré in trefoil, of 3. Notice what we can do. This is a trefoil that I've just placed on this table. Consider this particular crossing right here. Notice that this crossing is going on top of this other one. But, if I cut this open and now reglue it again, this crossing is going below, the strand is going under this one. By making that small change, I've actually changed my knot itself. What do I have now? Now I have just the unknot, so that 1 crossing change has changed my trefoil into my unknot.

But, remember in topology, I'm not allowed to cut and reglue because that has fundamentally changed the property of relative position. Let me ask you a very simple question. If somebody gives you the circle, which we're going to call the unknot in this particular projection, and somebody takes the circle, cuts it, ties it, makes a trefoil out of it, and gives this to you, are these both the same from the idea of a topologist? In other words, by isotope, by just stretching and pulling without cutting, can I make this trefoil knot into an unknot? Let's try it.

Here I am, I pull it, I twist it around, and I stretch it, and notice that no matter what happens, it doesn't seem like I can make the trefoil into the unknot. Our gut instinct says we have 2 very different objects topologically. No matter how much I stretch or pull I can't make one into the other one. But, a gut instinct is not a proof. It's just intuition. More generally we ask this question, are there knots? Are there knots at all? Is every piece of knot that we can come up with just the unknot in disguise? Is there a beautiful trick that takes this into the unknotted circle that we don't know yet? Maybe there's a special move that we have yet to learn that would make the trefoil that is actually knotted up, that we constructed by gluing the ends together into the unknot.

This is our goal for today, to prove that the trefoil is actually different than the unknot; that knots actually do exist. If we have this much of a hard time figuring out whether the trefoil and the unknot are the same or not, and if we only have instinct going for us, what about this picture? Here we see an example of a knot far more complicated than the trefoil or the unknot. Can we just look at this picture and understand whether it's the trefoil or the unknot or something far more complicated? What is it about this picture that will give us that information? Notice this picture has everything we need. It has the entire circle. It starts and ends at the same place. It has all the crossing information that you want. This has the information we need, but is there a way to tell whether this is the unknot or not? Before pushing ahead into mathematical language, let's consider how knots appear in nature. We're going to look at some snapshots today, but we're going to go into details of each of these in future lectures. Let me give you 3 quick ideas. First, knots appear in chemistry. In 1988, the first synthesis of a knotted molecule occurred. What a beautiful thing to think about, a molecule whose pieces are made up of atoms that come together and form a knotted structure. Remember what we talked about in the very first lecture, form and function are related, so the kind of knot that we get based on the molecular structure determines the property of the molecule itself.

They also appear in physics. Not just in chemistry in terms of knotted molecules, but in physics. A branch of physics called string theory believes that all matter is based on vibrations of circles. If you want to understand what the property of carbon atom is, it means there's an underlying circle that's vibrating that determines the property of that atom being carbon. If you want to understand what makes an atom oxygen, there's a vibration of a circle that does this. Not only is it the vibration of the circle, but it's the type of knots formed by the circle that determine the properties of this matter. Again, we see in physics that form and function are related, just like we saw in chemistry in the way these molecular knots are being built.

Third, knots also appear in biology. The DNA double helix is packed into the nucleus as a knotted coil. It's extremely long, this molecular structure, made up of thousands and thousands of atoms. But, it's packed into your nucleus. Certain enzymes are actually needed to go inside and untangle this knotted structure of the DNA, to smooth it out so your body can come in and use other properties to copy the DNA for cell division. Moreover, in biology, there's a type of fish called the hagfish, which ties itself up into a knot when a predator comes and tries to grab it. It then uses its slime that's around it and uses the knot theoretic properties to push itself away from the prey. The more we study these shapes, even as simple as knots, the more we push our understanding of nature.

The best way to study and work with knots is to not study the 3-dimensional objects that are knots, but look at their projections onto paper. We think of projections as their shadow. Imagine you have a knot, remember up to topology I can take this knot and stretch it and pull it and twist it around. As long as I don't cut and reglue, it's still the same knot. Instead of looking at this 3-dimensional object, note this is basically a 2-dimensional projection. If we just look at the flat projection of the knot, we can get the information of the knot we want. However, we need to draw in the crossing information; just looking at the shadow of the knot itself loses that crossing information. Thus, if I can redraw the crossing information again, then I actually have what we call the projection of the knot. It's the shadow of the knot with the crossing information that you need to rebuild this knot. For example, this picture we saw earlier is a knot projection. We see the knot itself flat on a 2-dimensional piece of paper along with that crossing information. The tools needed to study projections, which are basically pictures, which are basically shapes, was given by Kurt Reidemeister in the 1920s. Again, this is the start of the topology revolution. Reidemeister developed 3 moves—we're going to call it Reidemeister Moves I, II, and III for simplicity—which changes the knot projection, but not the knot it represents. In other words, these moves go into my knot projection, my knot picture, and they change the picture, but the knot itself is left unchanged.

Let me show you the first move to explain what I mean. Consider this picture. What this picture shows is a zoomed-in, magnified perspective of my small piece of my knot. My knot is all around this property, unchanged, and I've zoomed in and looked at this one piece of vertical strength. What Reidemeister says is his first move enables you to cut out that circle, leave the rest of that perfectly fixed to cut out that circle, and replace it with another circle with 1 little extra twist. Notice as a picture the knots are completely different in their diagrams, in their projections, because 1 had this vertical piece in it and the other one has the same vertical piece with a twist. We see the projections are different, but you can easily tell that the knot that it represents, either this one or this one with the extra twist, hasn't really changed at all. All I'm doing is a very small simple move in 3 dimensions, which hasn't changed my knot by cutting or breaking.

There are 2 other Reidemeister moves that we need to worry about. This first one says we can take this vertical string, twist it 1 way, or we can twist it the other way, and come back to the exact same knot, but have different projections. The other 2 are as follows. The second Reidemeister move says if you have 2 vertical lines, I can take one of these vertical lines and push it behind the other one, introducing 2 new crossings. Again, the knot projections are very different. They don't look like them at all. The pictures have changed, but the underlying knot has not changed at all. I've just pushed a strand behind another one in 3 dimensions. Similarly, the second part of the Reidemeister Move II takes the same 2 strands and pushes it on top. These moves are pushing it behind or on top is what the second Reidemeister moves are about.

The third Reidemeister move is as follows. I can take a crossing and take any strand that's behind the crossing and move it below that crossing. In other words, a crossing does not interfere with my strand that's behind it. Similarly I can take my crossing, take a strand in front of my crossing, and move it below it. A crossing at the strand completely above or completely below are independent of where they're placed. These are Reidemeister's 3 moves. The natural question is, who cares? Reidemeister came up with these 3 moves. Of course, by doing the moves, we have changed the pictures, but not the knot itself. It seems like you and I can come up with numerous other moves than just these 3. What's so special about these 3 moves? Why should we care about these 3 Reidemeister moves?

Reidemeister comes up with a remarkable theorem. He states that these 3 moves are the only 3 needed to go from one projection of a knot to any other projection of the same knot. In other words, given a picture, given a projection of 1 knot that you have and any other projection of the same knot, these 3 moves are enough to go from this to this. Of course, the way in which I performed the 3 moves, the combinations of which I do Reidemeister Moves I, then III, then II, then another II, could be complicated. But, he said regardless of how complicated you make the order sequence of the 3 moves, you will need nothing more than these 3.

Let's take a look at this picture. The question is, is this knot, which we call the trefoil, the same as this complicated knot right below it? Are they the same? I can find a collection of Reidemeister moves to go from one to the other one if they're the same knot. Consider this: I take my first trefoil, I do a Reidemeister Move I twist, then I push it through using Reidemeister Move II. Then, I take that crossing and go past that crossing; I take that strand and go past that crossing. We can do another twist, Reidemeister Move I, and I can take that twist and move it past under that crossing, another Reidemeister Move III. I've made the first one into my last one with the sequence of Reidemeister moves. Because of Reidemeister's theorem, we convert a global problem—something that shows that these 2 knots doesn't seem related at all, it's a global phenomena that 1 does not look like the other one—into a local problem. We just need to worry about small moves one at a time. This is the brilliance of Reidemeister's theorem.

The Reidemeister moves exist between 2 projections only if the knots are the same. We know by his theorem that if 2 knots are the same, then we can use a collection of Reidemeister moves to go from one to the other one. But, what if the knots are different? Try using a Reidemeister move for the trefoil and try to make it look like the unknot. Here's my unknot. Can I use Reidemeister moves, a Reidemeister move twist I, a Reidemeister Move III crossing, a twist and a crossing, can I just repeat these Reidemeister moves over and over again to try to make it look like the trefoil? If I fail after 100 attempts, how do I know that it's because I need to do 101 attempts? When do we stop? Maybe 1,000 attempts will eventually make this into my trefoil.

We see that these Reidemeister moves do not help us to tell knots apart. They do help us to tell knots that are the same. If we're given 2 projections of the same knot, we can go from one to the other one. But, if you're given 2 projections of possibly different knots, then no matter how much we try, we cannot make 1 into the other. How do we know if the trefoil is or isn't the unknot? It doesn't seem like Reidemeister has helped us answer this problem at all. He's given us a way to go between knots we know are the same, but not between knots which are different.

A major breakthrough using Reidemeister moves came in the idea of coloring. That's right, simple coloring that we learned way back in elementary school and preschool was developed and exploited in a beautiful way by a mathematician named Ralph Fox in the 1950s. Here is his definition. He said a projection of a knot is 3-colorable if it's 1 of 3 things we can satisfy. First, we can color every strand, which is a piece of a knot that goes from one crossing to another one. If we can color every strand using 1 of 3 colors, that's the first condition. The second condition is we need to use all 3 colors somewhere in my knot projection. Third, each crossing has to have all 3 colors meet at that crossing or only 1 color can meet at that crossing. That's his definition of 3-colorability. Note that 3-colorability is based on projections of knots. It's not based on the knot itself. If somebody gives me a knot like this, there's no way I can color strands. There is nothing called a strand in 3 dimensions because it's just 1 piece of rope. But, the moment I place it on the table and get a projection, now you have those over and under crossings that cut it up. The 3-colorability is based on projection, not the 3-dimensional knot itself.

Let's look at some examples and consider how 3-colorability works. Consider this picture of the unknot, this particular projection of the picture of the unknot. Notice here that this projection is not 3-colorable because I cannot use 3 colors to color the strands. But, what about this picture over here of the trefoil? Notice here this is 3-colorable. I can color 1 of the 3 different strands. Remember, where we got the word trefoil; I can color 1 of the different strands 1 of 3 colors. I can color this one red, this one blue, and this one green, and notice I've used all 3 colors. Each strand is 1 of the 3 colors and at every intersection 3 colors meet. The trefoil, this particular projection, is 3-colorable, whereas this particular projection of the unknot is not 3-colorable.

Let's consider another projection of the unknot, the 1 that we showed earlier, which was very close to that of the trefoil. What if we have this projection of the unknot? It's still the unknot. Note we can untangle this. Let's start coloring. If I start coloring the strands red, green, and blue, notice at this crossing all 3 colors meet, but at this crossing, and here only, 2 meet. I need to try to fix this. As I try to fix it, it turns out no matter how I color these different strands, I cannot succeed. This projection of the unknot is not 3-colorable. We proved one of the most remarkable results based on combining Reidemeister moves and 3-colorability. Here's the way we do it. If your knot projection is 3-colorable, we apply a Reidemeister move to say that it also keeps it 3-colorable. In other words, we want to understand how colorability and Reidemeister moves fit together. Let's look at an example.

Consider the Reidemeister Move I. Here we have a strand and notice that it's colored red; it's zoomed in to a picture of a more complicated knot. If I take this and apply Reidemeister Move I, if I twist it, I get a crossing. I'm going to color this entire strand red again. If the original projection was 3-colorable, notice that this new projection is also 3-colorable. Why is this? It's because here at this crossing either 1 color has to meet or all 3 meet. Notice 1 color meets, so it's also 3-colorable. That means that by doing Reidemeister Move I I've made it so that, if it originally was 3-colorable, doing a Reidemeister move still keeps it 3-colorable.

Let's consider Reidemeister Moves II and III, and see what this has to do with coloring. Consider this as Reidemeister Move II. Here we see that if I have 2 vertical strands both red, if I push one into the other one, either behind or above, I can color all of them red. By doing so, all the crossings satisfy my 3-colorability condition. You might be asking, aren't you supposed to use all 3 colors? But, notice, this is a snapshot of my knot. The rest of my knot could be using all 3 colors. At this particular spot, I know that if the original 2 vertical strands are 3-colorable, I can make this Reidemeister Move II also 3-colorable.

What happens if one of your strands is red but another one is blue, like this picture shows? Now what you do is, as you push the blue strand on top, you can color this extra new strand that appears to be green. Notice all my crossings are satisfying the condition of 3-colorability. Similarly, we see this is true for Reidemeister Move III. Here, I'm just going to give you 1 particular example. There are several other examples we need to check in terms of different colorings of strands. But, let's look at this particular one. Here I have 2 strands colored blue on the bottom and 2 red on top and a green strand going through it. If I use Reidemeister Move III and push that green strand down, notice what happens. I can now recolor that crossing red coming from the very top 2 strands, and the blues that are in the bottom I keep, and the green also stays the same. If it was 3-colorable on the left side, it stays 3-colorable on the right.

What does this have to do with our quest to tell knots apart? Recall Reidemeister's remarkable theorem. He says that every projection of a knot can be made by just these 3 Reidemeister moves. Let's combine these ideas of colorability that Fox gives us and this Reidemeister move given by Reidemeister himself, and here's what we get. According to Reidemeister, if 1 projection of the knot is 3-colorable, then every time we do a Reidemeister move it stays 3-colorable. If that projection is 3-colorable, I'm doing Reidemeister Moves, 1 after the other one, and it's staying 3-colorable. It's still 3-colorable, but Reidemeister says I could start at this projection and go to every possible projection there is, which means if this is 3-colorable, all projections of my knot are 3-colorable. If this is not 3-colorable, then I can't all of a sudden introduce a new crossing that it is 3-colorable, a new projection. If this was not 3-colorable, then every projection cannot be 3-colorable also. In other words, what we have found is that 3-colorability is not a property of the projection of the knot, but a property of the knot itself.

That's unbelievable because we didn't even know what 3-colorability meant for a knot. Remember we couldn't color a knot itself; 3-colorability can only be defined on projections. But, by this amazing result, 3-colorability turns out to be somehow a property that a knot itself has and it has nothing to do with its projections.

Remember the example we talked about earlier? Let's look at it again. Here we see this example with this 3-colorable trefoil knot at the top. Notice at each one of these steps, as I twist and as I push and as I do Reidemeister Moves I, II, and III, each one of my steps remains 3-colorable, which means that knots actually exist. Why is that? Let's think about our very first idea. Remember how the unknot, no matter how we colored it, no matter how we thought about the projections of the knot, became not 3-colorable? If that projection of the unknot was not 3-colorable, then the unknot itself does not have the property of 3-colorability. But, the trefoil, as we've just seen by this picture, is 3-colorable, which means every projection of the trefoil is 3-colorable. Thus, the trefoil has the property 3-colorability. Since the unknot doesn't have this property and the trefoil does have this property, they're different creatures altogether.

What have we done? We've proved a truly beautiful result based on shapes. It's based on Reidemeister moves, which all are local phenomena that control a global structure. It is based on colorings—simple ideas that we had when we were kids, applied in a powerful way. This is the type of creativity and originality that pushes the frontiers of math. Simple ideas such as color just cannot be discarded because it feels trivial. It has a powerful consequence of now telling knots apart. Thus far we are able to tell the simple unknot, the simplest of knots, and we have proven that this object and the simple shoestring knot, or the trefoil, are fundamentally different creatures. No matter what I do in terms of topology, of stretching and pulling, I can never make it into the unknot. I wasn't able to do this other than pure intuition, but now using the power of mathematics, we can actually understand shapes a little bit more.

Thank you so much and I hope you join me next time.

Creating New Knots from Old

Lecture 4

The fundamental problem in knot theory is to find stronger and stronger invariance. We want to tell apart more and more knots somehow, like a taste test.

This lecture begins by manipulating knots with the idea of addition. We can add 2 knots by making 2 cuts on the boundary of knot 1 and 2 on the boundary of knot 2, then gluing those cuts together to form a new knot. Knot addition is a small operation because we cut only on the boundary, and we introduce no new crossings. The new knot we have created is called a composite knot. We can also identify an "unknot," a concept similar to 0, that allows us to add a knot to it and get the same knot. We cannot, however, remove the complexity of knots using addition, as we can remove complexity with numbers. In other words, there is no subtraction of knots.

The notion of prime numbers can also be extended to knots. A composite knot is the sum of 2 nontrivial knots. A prime knot is not a composite knot; in other words, it can't be broken into 2 distinct pieces.

One goal in trying to tell knots apart is to find invariance. A knot invariant is a property, such as tricolorability, that is assigned to a knot and does not change as the knot is deformed. A knot invariant can distinguish between knots that are different but not those that are equivalent. The fundamental problem in knot theory is to find stronger invariants—to be able to distinguish among more knots. The dream is to find a characteristic for each distinct knot, which would be called a complete knot invariant.

The crossing number and the unknotting number are 2 classical knot invariants. The crossing number $c(K)$ assigns the least number of crossings that appear in any projection of the knot and is very difficult to find. A famous unsolved problem is to show whether the crossing numbers of 2 separate knots are related to the crossing number of the 2 knots added together.

The second classic knot invariant is the unknotting number $u(K)$, which is the least number of crossing changes in any projection needed to make the knot into the unknot. Just like the crossing number, the unknotting number is difficult to calculate, and it also has a famous unsolved problem: Is the unknotting number of one knot plus the unknotting number of a second knot equal to the unknotting number of the 2 knots put together?

This leap-frog relationship [between math and physics] is a fantastic way of research progressing, one motivating and pushing the other, like a big brother encouraging the younger brother, taking turns as to who's better.

Do the unknotting number and the crossing number both measure the same kind of complexity? Our intuition says that if we have a projection with the least number of crossings to draw, it's most likely the projection with the least number of strands to uncross, but that's not the case. Both numbers measure complexity, but they measure different kinds of complexity. When we're talking about the simplest projection of a knot, we have to know what we mean by simplest—is it the simplest to draw or the simplest to untangle? ■

Suggested Reading

Adams, *The Knot Book*.

Questions to Consider

1. In adding numbers, 3 + 4 = 4 + 3. Is this true for knots?

2. Can the unknotting number of a knot be greater than its crossing number? Why or why not?

Creating New Knots from Old

Lecture 4—Transcript

Welcome back and thanks for joining me again. Last time we began our study of shapes and we started with looking at these 1-dimensional objects called knots. We considered knots and their projections, their shadows on the floor, along with the crossing information, which led to the Reidemeister moves—those 3 great moves that Reidemeister said you can go from any crossing information of 1 projection to any crossing information of another projection using these 3 moves alone.

We then used the idea of 3-colorability on these knot projections and this helped us with our first true results, that knots actually exist. We were able to tell apart the simple circle, or the unknot, from the shoestring knot, the trefoil, using 3-colorability and using the Reidemeister moves in a beautiful way. We also showed how knots naturally appear in biology, chemistry, and physics. Today let's look at why physics should care about knots in a bit more detail. Last time we skimmed the surface of those 3 scientific areas and I just want to dig in a little bit deeper.

The physics revolution had a powerful impact on knot theory. Here's what happened from a big-picture setting. In the 1880s, Lord Kelvin, where we get the Kelvin temperature scale, considered knots to be the atomic model. This is what I mean by that. More complicated atoms were represented by more complicated knots. For example, if you have something as simple from an atomic structure, like hydrogen, which is a light atom, you would associate to it a very simple knot structure. Something as heavy as oxygen would get a much heavier, or more complicated, knot structure for it. This was Lord Kelvin's idea.

Several scientists, because of Kelvin's influence, started studying and classifying knots. They thought this was a groundbreaking field where these visually-complicated strings tied together gave an insight into how atomic structure itself worked. Interest in this area lasted for about 20 years of research and then Niels Bohr shattered it all. He and his group of scientists brought a completely different and better model of the atom which we use today. This model of the atom wasn't based on a knot structure, but it was based on electron orbitals and the nucleus of the atom which had protons and neutrons in it.

Soon, physicists completely started losing interest in knots. Since it really didn't capture the world in front of them and the world around them, then what's the point of studying knots in the first place anyway? But, mathematicians did not give up. Mathematicians study for the sake of understanding and not just for an immediate application to a real-world problem. We want to know what really is going on with these pieces of string and how we can tell them apart.

With the introduction of string theory, knots again became important to physicists in the past 20–30 years. As we talked about last time, string theory believes that everything—molecules, atoms, subatomic particles, quarks, bosons, everything—is governed by vibrations of super small strings, which are looped and knotted together.

One of the leaders in the field of string theory, a God in this world who has written over 300 research articles, is Edward Witten. He is the Einstein chair at the Institute of Advanced Studies in Princeton and he's also a Fields Medal winner. Remember a Fields Medal is given to the great mathematician and yet Edward Witten won this as a physicist. That's because he has revolutionized ideas that relate again the concept of strings back to subatomic particle models. We've come full circle. Lord Kelvin thought these strings somehow governed the way the universe works in terms of small objects, and Edward Witten and his group are thinking that this is probably true again.

This relationship of math and physics has always had a leap-frog relationship. Sir Isaac Newton, who although mathematicians would love to claim him as a mathematician, was really a physicist, introduced calculus and used it for his study on gravity. He needed a weapon to understand how gravity functions, how speed and velocity work, so he created a mathematical tool. Physics was already ahead of the curve and math needed to catch up.

Later on, Riemann brings in this idea of curvature of shapes, which we're going to talk about in future lectures. Riemann brought in the curvature of shapes as a mathematician, but Einstein later needed to use Riemann's ideas to talk about the curvature of space and time. Now mathematics took the lead. Witten and others have again brought in new math ideas in order to promote string theory. Now physics has taken the lead. This leap-frog relationship is a fantastic way of research progressing, 1 motivating and pushing the other, like a big brother encouraging the younger brother, taking turns as to who's better. It's fantastic. Now that we understand why physics cares about knots, let's actually jump in and do some mathematical way of thinking about these ideas.

The first part of this lecture focuses on manipulations of knots with the idea of addition. Remember what we talked about early on? We know how to add numbers and put numbers together, but what does it mean to add knots? Can knots be added together the way numbers can? Let's take a look at an example of what we mean by knot addition. Consider this following picture: Here you see Knot 1 and Knot 2. These are 2 distinct knots and the way I'm going to add them together is the following thing. I make 2 cuts on Knot 1 and I make 2 cuts on Knot 2, and throw out those extraneous strings that I've just cut out. These are very small cuts on the boundary, on the outside part of the knot. Then I glue those cuts together to get a new knot called Knot 1 plus Knot 2. Here is the most important thing about this knot addition. It's a very small operation because you only cut on the boundary at small pieces and moreover we introduce no new crossings.

Look at this picture again. If I had cut a small piece from the inside of 1 knot and a piece from the outside of another knot and now try to glue them together, notice that we have introduced these 2 new crossings; this is not the way we add knots. Anytime we put them together, no new crossings can be formed. This new knot that we have created is called a composite knot. It's made up of 2 pieces; it's composed of these 2 pieces. Remember, just like numbers have a special number called 0 where if you take 7 and add to 0 I get 7 again—is there something like this for knots? Is there a special knot that if we add to another knot we always get the original knot again?

Let's take a look at this picture. Here we see my first knot, but if I add to it the unknot, notice what happens. If I clip and glue it together, the unknot now becomes exactly the string that I had originally removed. I can just deform it, rubber sheet geometry, push it in and get that original knot again.

This procedure isn't just mathematically exciting, but it's actually crucial to string theoretic interactions in physics. This is how strings collide and new strings are formed.

How about subtraction? We talked about adding knots, but what about subtraction? Let me ask you this question: Is the unknot a composite knot? What does that mean? Is the unknot a composite knot? Let's take a look at this figure. Here we see 2 complicated knots, Knot 1 on the left and Knot 2 on the right. Is it possible to take 2 complicated knots, add them together, and miraculously get the unknot? Is this possible, that they cancel each other out perfectly to get the unknot? In other words, can we remove the complexity of knots by addition?

We could do this with numbers. If you're given the number 7, I could add to it a negative 7 and cancel out and get that unknot or the un-number 0. Is this possible for knots, though? In our guts, as we try to figure this out using real rope or real string, which I really encourage you to do, you will see that the complexity of this object always seems to increase. Thankfully this is not a proof. We actually need to do some work to get this. It turns out we need the power of 2-dimensional surfaces to prove that 1 knot added to another one will only increase complexity. In other words, the knottedness of knots cannot be removed like numbers. There is no subtraction of knots.

Just as numbers have the notion of prime numbers, numbers that cannot be broken down into different pieces, like 7 can only be broken down 7×1, we see that this notion can be extended to knots, and it's a fundamental idea. We define a composite knot as the sum of 2 nontrivial knots. If you can make a knot from 2 pieces that aren't obvious, we call it a composite knot. A prime knot is not a composite knot. In other words, you can't break the prime knot up into 2 distinct pieces. It cannot be pulled apart this way. Let's keep this in mind as we go through.

One goal in trying to tell knots apart is to find invariance. What is a knot invariant? A knot invariant is a property that is assigned to a knot, which does not change as the knot is deformed. As you pull the knot, as you stretch it, this property is fundamentally built into what the knot is rather than how the knot looks. In other words, it's a property that is assigned to a knot projection that does not change due to Reidemeister moves. If you have a knot projection and if you apply the Reidemeister moves to it, the invariant, what you've assigned to the knot, fundamentally should not change. It is something that belongs to the knot itself and it's not based on the way you draw the knot. It's not based on the picture of the knot.

In particular, remember last time we showed that 3-colorability is a knot invariant. It assigns the value yes or no to each knot. In other words, given a knot, if it's tricolorable we give it the yes value and give it another knot. If it's not tricolorable, we give it the no. No matter how you draw this knot or how you stretch or pull, the knot itself has a no characteristic of tricolorability no matter what you do.

A knot invariant can tell knots, which are not equivalent. If 2 knots are not equivalent at all, we can say this is a yes tricolorable, this is a no tricolorable. It can help us distinguish knots that are different, but it cannot tell us which knots are equivalent. Consider this following really silly example, but let's assign the value 1 to every knot in the universe. I claim that this is a knot invariant. Why would I claim that? It's because if I have a knot and I've assigned it the value 1, no matter how I push it or stretch it or how I look at the projection, it will always be 1 because it hasn't changed at all. I've given it the value 1 regardless of what I've done to it. This is an invariant; it does not vary as I pull and change the knot picture. However, this is a useless invariant. It's an invariant, but has no strength to it; it has no teeth to it.

The fundamental problem in knot theory is to find stronger and stronger invariance. We want to tell apart more and more knots somehow like a taste test. Tricolorability gives our tongue a sensitivity for 2 flavors. Is something tricolorable? Yes, it is. Is this tricolorable? No, it's not. That's all our tongue is strong enough to detect, yes or no. The dream is to find a flavor for each distinct knot, and we call this a complete knot invariant. Can we train our tongue mathematically to be able to taste each separate knot and have its own unique flavor? In this case, we can tell every knot apart from every other knot. Remember what tricolorability does. It only breaks up a world into 2 pieces. Yes it is or no it isn't. That's not a very sensitive tongue. We want to get more and more sensitive using mathematical tools. It turns out that this is one of the biggest unsolved problems in mathematics today. We cannot find a complete invariant.

Remember the issues with the trefoil and the unknot? No matter how many times we moved it around, we cannot prove that the trefoil and the unknot were different. We actually needed the power of an invariant to tell it apart, the invariant of tricolorability. Although it only breaks up into 2 pieces, it's at least strong enough to tell those 2 things apart. A strong invariant will have great uses in chemistry in terms of molecular design, in biology in DNA structure analysis, and in physics in string theory, like we talked about.

We begin by looking at 2 classical knot invariants: The crossing number and the unknotting number. It's a sense of comfort since it assigns numbers, something we're familiar with, to knots. We're associating a number to a knot. We already know how numbers work, so this is a good place for us to start. These 2 invariants can be thought of as different flavorings in your tongue. The crossing number can maybe taste saltiness and the unknotting number maybe sweetness. They're measuring different things, yet they give us a power to measure in the first place. This is what we want.

Let's talk about the crossing number first. The crossing number assigns the least number of crossings that appear in any projection of the knot. Let's take a look at this picture here. Note that this picture shows a projection of a knot with 6 crossings. What does the crossing number say? It says it is the least number of crossings over every possible projection. This picture has a knot with 6 crossings, which means the crossing number is 6 or less. We know this has 6, but maybe there's a way to move it around to decrease my number of crossings. Remember the crossing number is over every possible projection.

Notice if I do a simple Reidemeister Move I twist, I've decreased the crossing by 1. I get this new picture, which has crossing number 5 or less. I've already dropped my crossing by 1. Can I get it to 4 or less? If I do a Reidemeister Move II, I've knocked down 2 crossings at once. I get a crossing number of 3 or less. I know this knot, which happens to be the trefoil that we've seen before, is a crossing number of 3, but maybe we can draw it with 2 or smaller.

Here's what we know about crossing number. It's extremely difficult to find it. We know what the worst-case scenario is, you just draw a knot and you know what the crossing number's worst case is, but how do you get it to be lower and lower and find the smallest over every possible projection imaginable? Here's what mathematicians know so far for a glimpse of very low crossing numbers. Take a look at this picture. Here we see the trefoil, which is crossing number 3. We know that the trefoil can be drawn with 3 crossings and it cannot be drawn with less than 3 crossings. The figure 8 has crossing number 4 and these 2 other knots have 5 crossings each, and they don't have a name. We just call them 5-1 and 5-2. It's because knots get quite complicated as they go.

The main result about crossing number involves alternating knots, an extremely useful kind of knot. A knot is alternating if it has a projection where the crossings go over and under and over and under and rhythmically repeat as we walk around the knot. Consider the following picture. Here we see a knot, and notice as you walk around, it alternates crossings from over to under to over to under throughout the entire knot. We call a projection reduced if there is no easily-removed crossing. These 2 figures show examples of knot projections that are not reduced. In other words, if you look at the picture, is there an obvious crossing that you can simply twist and remove 1 crossing by this? It's basically a type I Reidemeister move feel. It's easy to check if a knot is reduced or not. We just look for these obviously simple crossings.

In 1986, 3 mathematicians independently proved a stunning result. They said that if you have a reduced projection, which is easy to check, and if the knot is alternating, which is easy to check, and if that picture has N crossings, then the knot itself must have crossing number N. In other words, there cannot be a projection with any fewer crossings no matter how we draw this knot. If you look at this figure we saw earlier, we see that this is an alternating knot, it's not reduced, it's the most simple way, there's no obvious twist we have, and it has 14 crossings. That means no matter what we do to this thing, it'll always have 14 crossings as the lowest number of crossings in any projection possible.

How do addition, which we talked about earlier, and crossing numbers interact? One of the most famous unsolved problems—and it's still open and it has been open for 100 years—is to show whether the crossing number of a knot plus the crossing number of another knot are related to the crossing number of the 2 knots added together.

Let's take a look at this picture. Here we see a knot, K_1—it's the trefoil in this case. We know it has crossing number 3. Here's another knot, K_2. It's the figure 8 knot; it has crossing number 4. What happens when I add these 2 knots? I get a new knot. I know that the crossing number of the first plus the crossing number of the second is 7, but what is the crossing number of this new knot that I've created? If I look at this picture, I see that it has 7 crossings, but can I move this picture around and make it drop crossings? Can I decrease crossings by doing this?

This open problem says that you can't, that the crossing number of 1 knot plus the crossing number of another knot equals the crossing number of the new knot you get by adding it together. But, mathematicians don't know how to prove it. We've tried it for every knot we can think of and it always works, but we don't know why. Isn't that a stunning fact, that we mathematicians no matter how amazing we are in proving deep and powerful theorems related to strings and chemistry and biology can't even prove a little result like this having to do with knots?

All the knot tables you will ever find that have classifications of all different knots always classify them based on prime knots, they're always prime, and they're always based on their crossing number, starting at the 3 crossing knot, the trefoil, and getting more and more complicated. Thus far in mathematics, we only understand knots with crossing numbers less than around 15 or 16; that's all we know. If somebody tells us to understand crossing numbers higher, we kind of run into a forefront of math research. This is what we want to use our weapons for, to push our understanding of complications of knots coming from crossing number.

The second classic knot invariant is called the unknotting number. Again, just like the crossing number, we're going to assign a number to a knot. But, this number isn't based on crossings; it's based on unknotting. Let me share with you a really beautiful result. If you give me the power to change any crossing from an overcrossing to an undercrossing or from an undercrossing to an overcrossing, if you give me the power to do this for any knot that can magically change crossings, then I can guarantee that I can unknot any knot you can think of. I can, in fact, take the most complicated knot and make it into the unknot, if you give me this power. Why is this true?

First, let's place our knot on a table and start at any point we want. Let's pretend I just lift my knot off the table at 1 strand and I'm going to walk along the table lifting this knot. If I come to an overcrossing, that's great; it's above anything else. I'm just going to lift it right up in space and keep it floating in 3 dimensions. If I come to an overcrossing, I just lift it up and keep it up here. If I come to an undercrossing, remember you've given me this power, so I'm going to switch crossings and make it an overcrossing and lift it up. As I walk along the knot, I keep lifting it up more and more into space. Notice what happens to the rest of the table. It's running out of crossings because most of my knot is up here floating in space. As I stretch and pull this knot and if you've given me the power of switching these crossings, I pull it up and eventually what I'd come up with is just an untangled piece of string, which is the unknot. Giving this power to me, I can untangle any knot, no matter how complicated, into the unknot.

What is the unknotting number? The unknotting number is called U of K, whereas the crossing number is called C of K. The unknotting number is the least number of crossing changes in any projection needed to make the knot into the unknot. The crossing number, remember, is the least number of crossings needed to draw the knot. The unknotting number is the least number of crossings needed to be switched to make it into the unknot.

Unknotting and knotting by changing crossings happens in nature all the time. DNA is the genetic blueprint of life, and nature uses an enzyme called the topoisomerase to actually manipulate the DNA, which cuts, changes crossing information, and reglues the DNA again. Why does nature need to perform this complicated operation? It needs to perform this because the DNA is crunched in into the nucleus, into the center of the cell, so you need something to untwist it, to open it up, to cut for replication and for other jobs you want to do with the DNA. Nature performs this unknotting operation of cutting and regluing and changing crossings all the time.

Similar to the crossing number, this is an easy invariant to define; I've just defined it. It's the least number of switches for any knot I need to make in order to make it the unknot. Just like the crossing number, it's hard to calculate. Consider the following knot. Here we see a knot, which has crossing number 8. To show that this is actually the crossing number is a difficult thing, but let's assume it has already been done. Consider these 2 crossings here and here. Notice that if I switch these 2 crossings from overcrossings to undercrossings and under to over, if I just change the crossings, I can actually untwist this entire knot and make into the unknot. I know its unknotting number is 2 or less. This only guarantees it's 2 or less. I know I don't need 3 because I already have 2 right here, but can I do it in 1? Can I unknot this thing with 1 twist? It was not proven until, amazingly, 1986 using some extremely powerful theorems that the unknotting number of this particular knot is actually 2.

Similar to the crossing number, the unknotting number also has a famous unsolved problem, which is: Is the unknotting number of the first knot plus the unknotting number of the second knot equal to the unknotting number of the 2 knots put together? Exactly like before, this is extremely easy to show in 1 direction, that we can show the unknotting number of the new knot is smaller than equal to the 2 put together, but we don't know how to show equality. Similar to crossing number, complexity of unknotting number seems to only increase when we start adding them together.

Let's close this lecture by bringing unknotting number and crossing number together. Does the unknotting number become realized in a projection with minimal number of crossings? Let me ask that question in a different way. Does the unknotting number and the crossing number both appear in the same projection of the knot? In other words, are they both measuring the same kind of complexity? They're measuring complexity; 1 shows it by complexity for drawing, which is the crossing number, and the other shows the complexity of unknotting, which is the unknotting number. Our intuition says, if you have a projection, which has the least number of crossings to draw, most likely that's the projection that's going to have the least number of strands to uncross. You only have a small number of crossings to work with, so that's a good place for us to start unknotting things.

In 1983, a knot was found, which has crossing number 10, where there's only 1 projection of this knot, which has 10 crossings. This is the picture right here. This knot has exactly 10 crossings and this is the only projection up to just reflection that has exactly 10 crossings. Given this projection, having only 10 crossings, it has unknotting number in 3 moves in this particular projection. If I switch 3 particular crossings in this projection, I can unknot this thing.

Let's look at this picture. Here is another projection of the same knot. This has 14 crossings, far more than before, but you can show looking at these 2 crossings right here that if I switch these 2 crossings alone, I can unknot this entire knot. I only need 2 moves now, not 3 like the last time. Therefore, although the unknotting and crossing number both measure complexity, they are measuring very different kinds of complexity. They're not as related as we think at all.

Both of these invariants, the crossing number and the knotting number, tell us different things about knots. It depends on which equivalents we care about. We can talk about the simplest projection of a knot. What do you mean by simplest? Do you mean the simplest to draw or the simplest to untangle? We have these 2 different ways of tasting knots, 2 different ways of measuring maybe saltiness, if that's what you're interested in, or sweetness.

We have come up with 2 classical invariants in knot theory. They're very easy to define and they're extremely hard to calculate. Our invariants are more powerful than 3-colorability, for example, able to tell more knots apart, but we can do better. We have also seen a deeper glimpse into physics and knot theory. In the next lecture, we're going to focus on biology and knot theory and consider multiple knots put together called links.

Stay tuned.

DNA Entanglement

Lecture 5

The structure of the DNA is extremely elegant, formed by the shape of this double helix with 2 strands perfectly interweaving between each other.

So far, we've looked at one circle knotted, but this lecture considers links, which are several circles knotted together. They can be individually knotted, or they can be tangled together in different complexities. We see, for example, the unlink (analogous to the unknot), the Hoff link (the link that forms chains), and the Whitehead link.

Similar to knots, links are equivalent up to isotopy. We can take the links in our hands and stretch them, but we're not allowed to pull or cut. For their projections, we again have Reidemeister moves. We can also identify links with more than 2 components, such as the Borromean rings. As it was with knots, one of our goals with links is to find link invariants—things based on the link itself that don't change. The number of components in a link is the first linking invariant. For example, the simple chain link has 3 components. The unlink, Hoff link, and Whitehead link have 2 components.

The linking number is a second invariant. To compute the linking number, we must first orient each component of the link by choosing a direction to travel. Next, we look at crossings between different components of the link. At every crossing, where 2 separate components meet, we give either a +1 value or a −1 value based on how we oriented the knot. We then add all the values at every crossing and divide by 2. Note that if we choose the orientations arbitrarily, the linking number values change. Thus, we must take the absolute value of our calculation. Note, too, that the Reidemeister moves do not change the linking number values.

We know from Reidemeister's theorem that we can get from any projection to any other projection with these 3 moves. That means that if we compute the linking number for this projection, we have computed the linking number for every projection possible and have found the linking number of the link itself. The linking number is an invariant.

How can we apply the linking number—this +/–1 crossing information—to knots? We call this calculation the writhe of a knot. We first orient the knot, then obtain a value at every crossing, and add the values. We don't need to divide by 2 because we don't have another component to worry about. And we don't need to take the absolute value because if we change the orientation, the writhe doesn't change. Reidemeister moves II and III don't change the writhe, but Reidemeister move I does, which means that the writhe is not a knot invariant. The writhe measures the kind of twist a belt goes through when a Reidemeister move I is performed on it. Thus, the writhe is not invariant for knots, but it is invariant for ribbons. Indeed, DNA can be thought of as a twisted ribbon in this helix.

This idea of adding up crossing information sounds beautiful, and it feels like we can use this beautiful idea on other things. Remember, we were motivated by this from the DNA linking. We naturally ask, why not try this idea on knots?

We have come full circle. We were motivated by looking at DNA with its double strand, which led to links. Then we then came up with the linking number, which was an invariant, and we pushed it on to knots to find the writhe of these ribbon strands. ■

Suggested Reading

Adams, *The Knot Book*.

Questions to Consider

1. Can you generalize the Borromean rings for four components so that if you remove any one component, all the other pieces become unlinked?

2. Is it possible to compute the linking number of a knot rather than a link?

DNA Entanglement

Lecture 5—Transcript

Welcome back and thanks for joining me again. A revolution has happened in the 20th century with an understanding of what the DNA structure looks like. The discovery of the shape of the DNA, that of the double helix, is one of the greatest advances of mankind. It was discovered by Watson and Crick in 1953 for which, rightfully so, they received the Nobel Prize. Indeed, the design of the DNA, this double helix, determined how it behaves as a molecule and how it interacts with other molecules. We see what we learned from the first of these lectures—that form and function are actually interrelated, even in the DNA as well.

The structure of the DNA is extremely elegant, formed by the shape of this double helix with 2 strands perfectly interweaving between each other. Let's try to think like a mathematician. What does a mathematician do with this? One thing you can do is take this idea and see how we can model this or generalize this principle of a double helix. From the world of a topologist, this twisting is what's important. Also important is the fact that there are these 2 strands weaving in forming the double helix. But, what happens when these strands, which weave beautifully in the DNA, are not so nicely organized? What happens if we throw in complexity? Instead of talking about 1 circle knotted, which we've been talking about so far, this lecture considers links, which are several circles that are knotted together. They can be individually knotted, in form of these links, or they can be tangled together in different complexities.

Let's consider some famous examples of links. Take a look at these pictures. Here we see the unlink. This is analogous to the unknot since there's no linking between the components and there's no linking within the components. The simplest form of link, where it's not the trivial unlink, is the Hopf link. This is the link we're most familiar with. These are the links that form chains. The Hopf link is simply 2 circles linked together in the simplest and most natural way possible. We also have the Whitehead link. The Whitehead link has a little extra twist compared to the Hopf link. Whitehead and Hopf are both very famous mathematicians where these links, based on their names, appear in very powerful ideas in 3-dimensional objects. We'll talk about some of those later. The Whitehead link also has a different presentation, a different projection, which looks like this. It's the exact same link except drawn a little bit differently, projected in a different way.

Similar to knots, links are equivalent up to isotopy. We can take the links in our hands and we can stretch them, but again, we're not allowed to glue or cut. For their projections, we again have Reidemeister Moves. Notice the projections of knots and links around a small area look identical. They're still pieces of string. Maybe the string is related to 2 different, separate components—that's where we get links from—or maybe they're strands of the same component, where we get knots from. Similarly, we can see that with this Whitehead link where we have these 2 separate projections. There is a collection of Reidemeister Moves, I, II, and III, that takes the left Whitehead link projection to the right Whitehead link projection.

Let's consider links with more than 2 components. These examples that I've given you so far are just 2 component links. What about the following? This looks somehow like a generalization of this Hopf link and it is. Here's a classic chain link with 3 components. The Borromean rings are based on an Italian family crest that appeared on their shields and their family logos. They have a beautiful property. Let me show you the Borromean rings and how they're different than this chain link here.

The Borromean rings look like this. You almost feel the Olympic symbol starting to be formed. Here's the beautiful property of the Borromean rings: if I delete any one of the rings, say the green one, what happens is that the red and the black are completely unlinked and I can pull them apart. Now, what happens if I delete the red one? If I delete the red one, notice that the green and the black can elegantly pull apart. Similarly, if I delete the black one, then the green and the red can pull apart. This is the power of Borromean rings. Although they're together, tangled together, separately, 2 by 2 by 2, they're completely trivial. They're completely unlinks.

This is not the case that we have with this chain link. If I delete that red component in the middle, of course, we can pull the other 2 apart. But, if

I delete the green or the black, you have this natural linking that goes on. Thus, the Borromean rings have this beautiful property.

Similar to knots, we are trying to tell links apart. From these 2 examples of links with 3 components, we see that 1 has these Borromean-like properties of pulling things apart by deleting links. But, we have to look deeper than just this. Thus, our goal is to find link invariance, things that don't change based on the link itself. No matter how we look at the projections of the link, we still want to end up with the same property associated to it.

Remember, an invariant is a property that does not change as our links are deformed. No matter how I pull or stretch, the fundamental property of an invariant is based on the link itself and not the way we look at it, the way we stretch it, or the way we project it and look at their projections. The number of components in our link is our first linking invariant. This is completely easy and the most natural thing to think about. For example, the simple chain link has 3 components. The unlink, Hopf link, and Whitehead link, as these pictures show, have 2 components. Thus, the chain link and these different links are all different based on the number of components. The unlink has 2 components. The Hopf link has 2 and the Borromean rings have 3. Therefore, we can tell those apart because of the number of components. That seems quite easy. We want more, we want more.

Under this invariant of number of components, notice that all knots get the value 1. Thus, this invariant, although it is an honest invariant, tells the knots apart in no way possible. All knots get grouped into 1 clump. The linking number is what we're interested in today. The linking number, which is between any 2 components of our link, is our second invariant that we care about. The first is the number of components and now we're going to find a new way of calculating this number.

The linking number is computed as follows, but before we go into the calculation, I want to make you understand what it's really about. Let me show you this picture right here. This is a sample of 2 links put together. You see the complexity of the white strand and the complexity of the black strand. What I'm really interested in is not the complexity of the white or the black, I'm just really interested in how they're related to one another.

In other words, I don't care about the complexity of the white or about the complexity of the black, just their linking between them. Let's find a way to measure this. Let's take a look at some pictures. The first thing we do in order to compute the linking number is we do something called orientation. We orient each component of the link by choosing a direction to travel.

Consider these 2 components. I pick any place I want and I arbitrarily choose a direction to travel. You just draw a simple arrow telling you that's the direction you're going to travel throughout this entire component. Similarly, I can pick an arrow on the other component completely independently. I'm going to pick this arrow for both of these guys and this is the way they're going to travel throughout my link. Great! That's the first thing we do.

The second thing we do is we look at crossings between different components. Remember what we talked about last time. We're not interested in how 1 component mixes with itself. We want to know the crossings between these 2 components are put together. We only focus on crossings of 2 separate components, where the red component and the black component meet. We look at each crossing such that the arrows of that crossing, in a zoomed in perspective, are always pointing towards the top.

Let's take a look at this picture. When I zoom in close enough to my knot picture, to my link picture, what I want is for the 2 arrows to be pointing towards the top. What if they're not pointing towards the top? I can always rotate my head or move my link around so that, according to the orientations, my arrows are always pointing to the top. If my strand with 2 arrows pointing to the top—the strand that's going on top, the overstrand—has a positive slope, rise over run is a positive value. Then, I assign a +1 number to this crossing. But, if my crossing, with the 2 arrows pointing to the top again, if my overstrand at this crossing has a negative slope, then I assign a −1 to this crossing.

At every crossing, where 2 separate components meet, I either give a +1 value or a −1 value based on how I oriented the knot in the first place. Here is what we do. We take all the values that we get for the entire link, at every crossing where 2 separate components meet, and we add them together. For example, in this previous picture, we see 1, 2, 3, 4, 5, 6 places where these

2 red and black components meet. Each 1 gets a value of +1. We add it all together, we get the number 6, and then we do something very important. We divide by 2. Since we have these 2 components that meet, we divide it by 2. Thus, the value of the linking number, of this particular link, is 3—6 divided by 2. Some of you might be thinking, this linking number seems to be based on the arrows and how its pointing, that's how I get the ±1 values. What happens if you pick your orientation the way you want to travel around the components, and I pick my orientation the way I want? Will this value change?

take a look. What happens if I take my crossing of 2 different components, with both arrows pointing up? Here, the red has that overcrossing, so my value at this crossing is +1. What happens if I chose to orient my red component in the opposite direction? If I chose my red in the opposite direction, my arrow would now point down instead of up. Remember how we always measure the ±1 values, they always have to be pointing up, so I need to rotate my head, or I need to rotate the picture to make sure those 2 arrows are now pointing up. When I do that, I notice the following thing: What used to be a +1 crossing—because the red had that positive slope—now since my red's arrow has switched—now my red has a negative slope. All my +1 values become −1. Similarly, if I look at this second piece of the puzzle, were I have my red behind my black, where my black has this negative slope, if I switch my red values around to point the other way—now I need to rotate my head to make sure I'm pointing my arrows in the same way to measure my value—what used to be a −1 because my black has a negative slope, now becomes +1 because now my black has a positive slope.

What does this mean? This means if I choose the orientations arbitrarily, my linking number values change. Thus, we want to take the absolute value of the linking number. This way, once you take all your values of plus and minus ones in your crossings, add them all up, and divide by 2, we take the absolute value and we don't have to worry about the way we oriented it. Now we come to a natural question. What does this have to do with an invariant? It seems like I just made it up. Why is this necessarily an invariant? Remember, we need to check the 3 Reidemeister Moves to make sure that this value of the crossings that we're getting, this linking number, does not change based on the 3 Reidemeister Moves. Let's take a look.

Look at this first figure. Here you see Reidemeister Move I. If you look at Reidemeister Move I, I introduce a vertical line, which now, because of a twist, has an extra crossing here. But, if you notice what the linking number measures, it only measures crossings between 2 separate components. Thus, in this particular picture, that new crossing that I get does not even register for my linking number because it's a crossing of the same component with itself. What used to have linking number value, which contributed nothing before, now with this extra twist still contributes nothing. I get no new values from my linking number to change based on Reidemeister Move I. I can perform Reidemeister Move Is all I want and my linking number does not change.

What about Reidemeister Moves II and III? Let's take a look at these. If you have Reidemeister Move II, let's orient each of these 2 strands up. Have one of the black ones pointing up and let's have one of the red ones pointing up. On the left side, before I perform my Reidemeister Move II, how much does this contribute to my linking number? You see that the strands don't even cross at all. Therefore, my linking number does not get a new value based on this local property. It has net contribution of 0.

What happens if I take my red strand and push it under my black strand? I've introduced 2 new crossings. Let's look at these crossings in detail. Remember, I've arbitrarily oriented my 2 strands to be pointing up. If I zoom in at these crossings, you see that the first crossing, the 1 on the bottom, has a value of $+1$. The top crossing has a value of -1. Thus, the net contribution that this would give to my linking number would give me a $+1$ and -1 value—it would give me a value of nothing extra.

My Reidemeister II Move—which used to have a value of nothing extra contributed to my linking number—performing this Reidemeister II, 2 crossings are introduced, but they both cancel out perfectly. My linking number value does not change as I perform Reidemeister Move II. What about Reidemeister Move III? Let's take a look at an example. We can color these 3 strands in several ways, but I've chosen to color 1 black and the other

2 red. There are 3 crossings here. We only care about 2 of them because the third crossing has 2 red components, the same component, so we just leave it alone.

On the left side, if I orient all my strands arbitrarily, and make them point to the top for now, then we see at the bottom, we have a value of -1 coming to my linking number. On the top, we have a value of $+1$. Thus, if this was my local picture that I'm taking the snapshot of, my linking value at the total gives me a net value of 0. My linking value, so far, is unchanged.

What happens if I take that black strand and cross it over that crossing, the 2 red strand crossings? Now if I look at the bottom crossing, which used to be a -1, is now a $+1$. The top crossing, which used to be a $+1$, now becomes a -1. This means that if I perform Reidemeister Move III, my crossings, although the values are different, the total value remains the same. There is no change.

What have we learned? The Reidemeister moves do not change the linking number values. If I do Reidemeister Move I on a particular projection, nothing changes. If I do Reidemeister Move II on a particular projection, its values don't change. If I move Reidemeister Move III on a particular projection, it doesn't change again. We know from Reidemeister's theorem that I can get from any projection to any other projection with these 3 moves. That means if I compute the linking number for this projection, I have basically computed the linking number for every projection possible. This means I have found the linking number of the link itself. The linking number is an invariant.

Let's consider some examples of the linking number to actually calculate this. Here we see the linking number of 2 circles which aren't touching each other, the unlink, to be 0. There's no crossing and thus it has linking number 0, which is intuitively obvious for what we want it to be. Fantastic! What about the linking number of a Hopf link? We orient any way we want. Remember, at the end, we're going to take the absolute value so it doesn't matter. We orient any way we want. We look at the 2 crossings. Both crossings have, in this particular picture, a value of $+1$. Thus, I add them together; I get $+2$, divided by 2. The Hopf link is linking number 1. This makes sense because it's the simplest form of linking we can think of. This is why we divide by 2—because this link, the simplest link, you want to have a linking number 1.

What about the Whitehead link? If we look at this particular picture, we have 4 crossings. Two of them, the right 2, turn out to be +1 and the left 2, both crossings, have value −1. I add up all 4 things and I get total value of 0. That's what the linking number says, which means no matter how I move the Hopf link around, I'm going to get value 0. But, didn't I have the value 0 for the unlink? The linking number is an invariant, but it's not that powerful. It is able to tell the Hopf link apart from the unlink, but it cannot tell the Whitehead link apart from the unlink. Just like knots, we want to be more clever in trying to find ways of studying the linking number.

How does a mathematician think? This idea of adding up crossing information sounds beautiful and it feels like we can use this beautiful idea on other things. Remember, we were motivated by this from the DNA linking. We naturally ask, why not try this idea on knots? It seems like, so far, all we have tried on knots are coloring. Maybe we can take this plus/minus 1 crossing information, throw it on knots, and see what we can get out of it. Let's see what we can salvage.

We now introduce a way of trying to make this work on knots. We call this calculation the writhe of a knot. Here's what we do. The first thing we do, just like with links, is we orient the knot. The second thing we do is obtain a value at every crossing. Remember, the linking number only cared about values between 2 separate components. It didn't care about the same component. Here, we only have 1 component, so we need to be sensitive towards everything. There's no other component for us to worry about because it's a knot.

I look at each crossing and I give a value to it. In this particular case, we give a crossing value of +1, −1, −1, −1, +1, +1, and +1. Those are all my crossing informations based on the identical way I used to do it before. Point the arrows to the top. If you have a positive slope, you get +1. If you have negative slope, you get −1. What do I do? I add all the values up just like before and I get the value +1. There are 4 +1 values in this link and 3 −1s, so my sum becomes +1. Here, we don't need to divide by 2; we just leave this value +1 here. We don't need to divide by 2 because there is not this other component I need to worry about. It's the same component and this is how I find the writhe. You might say, what happens if I change the crossing information based on the orientation? What if I orient to this exact same knot in a different way? Maybe I would get a different value for the writhe. What about the absolute value that I did before? Let's take a look at this example.

If I take this crossing information right here and change the arrows because of my orientation, both my arrows change. Why? It's because my entire knot changes in terms of its orientation. This crossing arrow becomes down and this crossing arrow also points down. Instead of rotating my head 90 degrees, I have to rotate it 180 degrees to bring the arrows up again. When I do it, I see a crossing that used to be positive remains positive. A crossing that used to be negative remains negative, which means if I change orientation, my writhe doesn't change.

Let's consider some really simple examples. Here we have the trefoil. If we compute the writhe of the trefoil, we see that you have 3 −1 values and we get a −3. Great! If we compute the writhe for the figure 8, we have 4 values—2 +1s and 2 −1s. Remember, orientation doesn't matter, writhe does not care. At the end of the day, we get a value of 0. You might ask the question, who cares? Is this just a silly calculation or can we actually say something about this and make it used for knots? Remember, last time we proved that the linking number was an invariant by doing the Reidemeister moves. It became a useful, powerful tool; I've just made up this rule. Mathematicians are trying to see what we can get out of it. We make up this rule for knots and see if it's an invariance.

Let's take a look. Is it an invariant under Reidemeister Moves II and II? If we take a look at Reidemeister Move II, we see that we have 2 of the same strands just like before. If I cross under, I get a +1 and a −1 value. This is exactly what we did last time for the linking number. All my proof of my Reidemeister II for the linking number is identical for my writhe. Fantastic!

What about my Reidemeister Move III? I have 3 crossings and now I have that middle crossing to worry about. I have a +1, a +1, and a −1. I do my swing over of my strand on the other side and that central crossing still remains +1. The other 2 now become −1 and +1. They switch just like before. In fact, I'm not doing anything new. My total writhe on 1 side added up to +1 and my total writhe on the other side adds up to +1, which is great. The Reidemeister Move II works and the Reidemeister Move III works. But, what about Reidemeister Move I?

We've saved the best for last. Let's try it. If I take a Reidemeister Move I, notice that it gets no crossings, so it has no contribution towards my writhe. But, if I put a little twist in there, look what happens. My writhe value increases. If I have a Reidemeister Move I and I put in another kind of twist, the opposite twist in my R_1 moves, my Reidemeister Move I, it decreases. This is depressing. This is not a knot invariant. It works for Reidemeister Moves II and III, but it doesn't keep the same values for Reidemeister Move I.

What does a mathematician do? We say, this is not a bad thing at all. Somehow what we have created is sensitive to Reidemeister Move I. It doesn't care about the II and III, but somehow it can taste that Reidemeister Move oneness. What does this number measure for knots? It's not a knot invariant, but it's measuring something.

The answer is that it's measuring something to do with my belt. Let's pretend I have a collection of belts here and I perform Reidemeister Move II crossing over those belts or Reidemeister Move IIIs. Notice that my belt itself hasn't changed shape as I perform crossing changes with other belts that are lying right next to it. Let's perform a Reidemeister Move I. To perform a Reidemeister Move I, you have a move that looks like this. If I pull it, notice that my belt itself has fundamentally twisted. This is what the writhe measures. It's measuring the kind of twist a belt goes through when you perform a Reidemeister Move I.

This belt does not go through any kind of a twisting by II and III because I'm just pushing the belt around over the crossings. However, if I do a Reidemeister Move I this way, it twists in 1 direction. If I perform a Reidemeister Move the other way, it twists in another direction. We see this kind of twisting is exactly what the Reidemeister Move I measures. The writhe is not an invariant for knots; it is an invariant for ribbons. Imagine you have a ribbon or a belt that forms in the shape of a knot and the writhe is useful for measuring such ribbon knots. Indeed, the DNA itself can be thought of as a twisted ribbon in this helix. This is what the Reidemeister move is measuring in terms of the writhe.

We have come full circle. We were motivated by DNA with a double strand. This led to links. Then we then came up with the linking number, which was an invariant, and we pushed it on to knots to find the writhe of these ribbon strands.

In conclusion, what have we done? We've considered DNA motivating our work, which has helped us to generalize to this idea of a writhe. The writhe itself is not a knot invariant, but it is useful for thickened strands that look like ribbons. This is a general property of mathematics. We take an idea like the linking number and we prove that it works. We generalize and extend and push it on to knots. We try to salvage it as much as we can and if we can't, we notice what it really is trying to measure.

In the next lecture, we study one of the most powerful knot invariants ever discovered, the Jones polynomial.

Stay tuned.

The Jones Revolution

Lecture 6

How powerful is this polynomial? A big, open question has been ... can we tell any knot from the unknot? Any time we use the Jones polynomial, we get the value of the unknot as 1. But any time we use the Jones polynomial of any other knot ... the Jones polynomial is not 1. It seems that the Jones polynomial is able to tell every knot apart from the unknot.

In this lecture, we will turn to the power of algebra, which measures structures in the world of topology, and study a new algebraic knot invariant. This invariant does not assign a number or a property to a knot. Instead, it assigns a polynomial. This polynomial is called the Jones polynomial after its discoverer, **Vaughan Jones**, who found it in 1984.

A polynomial, such as $5a^3 + 4a + 2$, can actually be thought of as simply a set of numbers. In this case, the set is 5,0,4,2; 5 for the amount of a^3, 0 for the amount of a^2, 4 for the amount of a, and 2 for the amount of the constant that is not a. With this in mind, we will build the Jones polynomial (the bracket polynomial), using a method developed by Lou Kauffman.

The bracket polynomial is based on 3 rules. The first rule tells us that the bracket value of the circle, which in the projection we see is the unknot with no crossings, equals 1. The second rule tells us the relationship between the 3 polynomial values. Whenever we get one polynomial, we can multiply it by a of the other polynomial plus b times the third polynomial, and the equality will work. We can also think about the second rule as follows: If we have a positive slope, we cut vertically and horizontally; if we have a negative slope, we cut horizontally first, then vertically. The third rule gives us another relationship, that between a link with an extra circle and the polynomial of just the link.

If we're given a knot or a link and we use the second rule repeatedly, it keeps removing all our crossings and makes vertical and horizontal cuts. If we keep applying the rule, we are left with a collection of circles because all the crossings are gone. Then, the third rule says that every time we have a collection with a free-floating circle, we can throw it away as long as we have multiplication by c to the polynomial. The second rule destroys all the crossings into circles; the third rule gets rid of the circles; and the first rule says that if we're left with one circle at the end, its value is 1.

Edward Witten, … a superstar physicist, who himself won a Fields Medal, related the work of the Jones polynomial that Vaughan Jones came up with to work in string theory and in 3-dimensional objects.

Our goal is for these rules to satisfy some kind of knot invariant properties. We want to make sure that the polynomial does not change as we perform Reidemeister moves. With a great deal of algebraic manipulation, we see that the polynomial is invariant for Reidemeister move II. Moreover, Reidemeister move III emerges easily out of the work we performed for move II. Unfortunately, Reidemeister move I fails us again, just as it did when we were working with the writhe. However, we can combine both the negative properties that cause Reidemeister move I to fail and make the invariance work.

Any time we use the Jones polynomial, we get the value of the unknot as 1. Any time we use the Jones polynomial of any knot that is not the unknot, the value is not 1. Although it's still an open question, it seems that the powerful Jones polynomial is able to tell every knot apart from the unknot. ■

Names to Know

Jones, Vaughan (1952–): Winner of the Fields Medal in 1990, he created one of the most powerful knot invariants.

Witten, Edward (1951–): A mathematical powerhouse who received the Fields Medal in 1990, Witten is considered the greatest physicist of our time, known for his work in string theory.

Suggested Reading

Adams, *The Knot Book.*

Questions to Consider

1. Compute the Jones polynomial of a knot of your choice.

2. How would things change if we defined the bracket <0> of the circle to be a value other than 1?

The Jones Revolution

Lecture 6—Transcript

Welcome back and thanks for joining me again. So far, we have been using colors and numbers in trying to tell apart knots and links. We talked about 3-colorability. We talked about the linking number. We even talked about the writhe. Not only is our pursuit of invariance useful for biology, chemistry, and physics, but it also gives us insight into how mathematicians struggle to attack problems.

Remember, so far, our invariants have been focused mostly on numbers. The crossing and unknotting number that we came up with were easy to define, but they were really hard to calculate. It took some powerful theorems beyond the scope of these lectures to actually show that certain knots have certain crossing numbers. Similarly, linking and writhe were also numbers. They were a bit harder to define than the crossing and the knotting number, but they were quite easy to calculate. Somehow we have to pay the price—whether it's easy to define or easy to calculate.

Today you're in for a real treat. We introduce one of the most powerful tools available to us, that of algebra. We're going to bring in the power of algebra, which measures structures in this world of topology. A new algebraic knot invariant is going to be studied. This knot invariant was discovered in 1984. It shook not just the mathematics community, but it also shook the physics community as well. In 1984, Vaughan Jones, while working on an unrelated area of mathematics, stumbled upon an invariant for knots.

This invariant does not assign a number to a knot, nor does it assign a yes/no value of whether it is or isn't 3-colorable or some other property. It assigns an entire polynomial to a knot. Because of this discovery, Jones was awarded the Fields Medal for his work. Remember, the Fields Medal is the greatest honor that could possibly be given to a mathematician who is younger than 40. Edward Witten, who we talked about last time as a superstar physicist, who himself won a Fields Medal, related the work of the Jones polynomial that Vaughan Jones came up with to work in string theory and in 3-dimensional objects.

This polynomial isn't just great for mathematics as coming up with a beautiful invariant. It actually pushes the frontier into science as well, especially physics. Currently, there are numerous polynomial invariants. Remember, I'm telling you something that happened in the 1980s. Several have come about due to the Jones revolution, generalizing the work of Vaughan Jones and several others. In 1928, many, many years before Jones' discovery, Alexander actually found an invariant, which assigned a polynomial to each knot.

However, unlike the Jones polynomial, it was not related to ideas outside of mathematics. Thus, the revolution didn't really take place anywhere other than in the topological circles. We're going to focus on the Jones polynomial rather than the Alexander ones or rather than future generalizations of the Jones polynomial that came up after the 80s. We're going to focus on the Jones polynomial due to its historical significance and its power.

I'm going to give you a word of warning about today's lecture. For the previous invariants, I've simply told you how to find them. I've told you how to compute tricolorability—color the strands and look at the crossings. I've told you how to come up with the linking number—add +1 or −1 values depending on the slope. Today, we're going to build this invariant from scratch in order to motivate a deeper understanding. Moreover, it's going to give us a glimpse of how mathematicians think and struggle with problems, the way we got a glimpse of this in terms of how the writhe worked.

When things failed, you pushed through and you came up with a new way of thinking about what Reidemeister Move I is really about. It's like my wife's homemade blueberry pie with her special homemade crust. It's more painful, but it's worth it. It turns out that it's actually more painful for her and worth it for me, but that's a separate story.

First of all, what is a polynomial? We have seen examples of polynomials all our lives, such as $5a^3 + 4a + 2$ or $3a + 5 + 8^{-1}$. Some are more comfortable with using the letter x instead of a. For example, $5a^3 + 4a + 2$ could be $5x^3 + 4x + 2$, but this is just placeholder information.

We're going to just use A throughout these lectures. You can think of them as X if you're more comfortable. These polynomials, such as $5a^3 + 4a + 2$, can actually be thought of as simply a set of numbers. Notice that the a isn't doing anything amazing. It's keeping the place for where that constant in front of the polynomial is. Therefore, $5a^3 + 4a + 2$ can be thought of as 5,0,4, 2; 5 for the amount of a^3, 0 for the amount of a^2 there is none, 4 for the amount of a, and 2 for the amount of the constant which is not *a*s. Similarly, $3a + 5 + a^{-1}$ can be thought of as the collection of numbers, 3, 5, and 1. This is because 3 keeps track of the a, 5 keeps track of the constant, no *a*s, and 1 keeps track of how many a inverses, a to the -1s there are.

Again, it doesn't matter whether you use *a*s or *x*s. I want us to think of polynomials not just as numbers, but as a collection of numbers. Instead of assigning a yes or no to a knot, like 3-colorability, or assigning a number like the crossing number, the unknotting number, or the writhe to a knot, we assign a set of numbers, a polynomial's worth. We're going to show that the construction of the Jones polynomial using a method created by Lou Kauffman. This is called the bracket polynomial. Remember what our motivation is. We're going to build this from scratch and this way of building the polynomial from the beginning was given to us by Lou Kauffman.

This bracket polynomial, our motivating polynomial, is based on 3 rules. Let's take a look at these rules. Rule number 1 says that the bracket value of the circle equals 1. Note that I am not talking about a particular projection of the unknot in complicated ways, but this particular projection itself. In other words, the bracket of the circle, which is the unknot with no crossings in this particular projection, gets value 1. That's the first of my 3 rules that I need to build this polynomial up. If you're given the bracket polynomial of a more complicated way of drawing the unknot, rule number 1 has nothing to do with it. It says, I don't know what to do, all I know is, if you give me the circle projection of the unknot, it's going to be 1. It's great.

Rule number 2 says the following thing. If I have a crossing as follows, I can replace this crossing of the bracket polynomial of that picture by A some polynomial value times this particular way of splitting the crossing vertically plus B times this way of splitting the crossing horizontally. What does this mean? Let's take a look at this picture here to understand.

Imagine you are given this trefoil and imagine we want to know what the bracket polynomial of the trefoil is. Rule number 2 does not tell us what it is. It doesn't just give us the answer like how to compute the linking number. It tells us a relationship between the 3 polynomial values. For example, it says the bracket polynomial of this particular trefoil projection equals A times the bracket polynomial of the trefoil where we replace one of those crossings with a vertical cut plus B times the bracket polynomial if we replaced one of those crossings with a horizontal cut.

Now, it has a relationship based on this rule with how the trefoil—whatever the bracket polynomial of the trefoil is, which we don't know—has a relationship between the bracket polynomial of the trefoil to A times the bracket polynomial of this knot diagram plus B times the bracket polynomial of this link diagram. It simply gives us a relationship. I want us to remember that, although it doesn't tell us the answer, we're building towards getting this answer. Moreover, consider the middle part of this second rule example that I've given you.

Notice here that when I make this vertical cut, I actually come up with the unknot. But, this is not the unknot where the projection is just a circle. Thus, what I have here is the bracket polynomial of the unknot of a different projection of the unknot, not the simple one. Again, I still don't know what that value is. It's just a relationship between these 3 separate polynomials. The moment you get 1 polynomial, you can multiply it by A of the other polynomial plus B times the third polynomial and the equality works. That's what the second rule is.

If you go back to the second rule again, you notice that there's another way of thinking about this exact same rule. If I take my positive slope crossing and cut it vertically to get A times that vertical move plus B times a horizontal move, then if I got a negative slope crossing. This is the same thing as A times a horizontal cut plus B times a vertical cut. Rule 2 says that, if you have a positive slope, you cut vertically and horizontally; if you have a negative slope, you consider a cut of horizontal first and then a vertical.

What does rule 3 say? Let's take a look. It's actually an elegant rule. Rule 3 says the following thing. If you take a link, no matter how complicated it is, it could be a knot. If you take any picture you have of a diagram and along with that picture if you just have a separate unknot that's purely a circle next to it then the bracket polynomial—whatever that polynomial is, which we don't know yet—of this picture is the same as the bracket polynomial of just the link alone. In other words, I can take the circle and throw it away as long as I multiply this polynomial by C, another variable. This is what we know about rule 3. Notice it's another relationship. It gives a relationship between a link with an extra circle along with the polynomial that you would get of just the link.

As we start using rule 2 repeatedly, what happens? Remember what rule 2 does. It takes a crossing and it cuts it vertically and horizontally. If you're given a knot or a link, as we start using rule 2 repeatedly, it keeps removing all our crossings and makes vertical cuts and horizontal cuts. Our crossings start disappearing. We get a relationship of these other components, but they have less crossings than I started with. At the end of the day, if we keep doing this, we are left with a collection of circles because all my crossings are gone. I just have a collection of circles.

What does rule 3 say? Rule 3 says, every time you have a collection with a free floating circle you can throw it away as long as you have multiplication by C to the polynomial. Rule 2 destroys all my crossings into circles. Rule 3 keeps getting rid of my circles. I can keep throwing it away as long as I multiply by the C. Rule 1 says if you're left with 1 circle at the end, which we're going to be, the value is 1. But, during this entire step we have manipulated and changed our polynomial based on crossing information, circle removal, and leaving it with that last circle. We believe that this is enough for us to actually get this polynomial.

So far, this is absolute nonsense because I'm just giving you 3 rules. What does this have to do with anything? We want these rules to satisfy some kind of knot invariant properties. As we struggle with the linking number and the writhe last time, we want to make sure that these rules are enough to make sure that the Reidemeister move values don't change as I change the projection of my knot. Note that the bracket value depends on the particular fixed projection. We cannot simply move it around. Remember rule number 1, it only works on that particular circle.

I want to make sure as I'm performing Reidemeister moves that my polynomial does not change. Let's see what this implies for Reidemeister Move II. Let's start with Reidemeister Move II. Look at this picture. In Reidemeister Move II, we have these 2 vertical strands that are crossing. What can I do using my rules I, II, and III? I'm going to look at the top crossing. Notice it's a negative slope crossing, which means I take this crossing, cut it horizontally first, and multiply it by A, plus vertically first and multiply it by B. This bracket polynomial of this particular knot, whatever happens to be with this crossing information of Reidemeister II zoomed in is A times this picture plus B times this picture.

Let's zoom in on each of these pictures and study it separately. Look at the first picture, that first diagram that's associated to the 1 next to A, that polynomial. This has an additional crossing. I'm going to look at this crossing; it's in positive crossing. I take this crossing, cut it vertically, open it up, and multiply by A. Then I cut it horizontally and multiply by B. I'm doing rule 2 again. Now you have these new pictures—A times this picture plus B times this picture. The A picture can simply be reduced without changing crossing, just stretching to make it look like these 2 horizontal strands.

But, the second 1 has this extra circle. I can just throw that circle away as long as I multiply my answer by C. Here's what I get when I look at this first part that's associated with A in the very beginning of my Reidemeister II Move.

What happens to the second part, the 1 that has associated with B in my very beginning Reidemeister II Moves? Let's zoom in on this. If I have this bracket, I can look at this crossing, it's a positive crossing. I cut it vertically, A times this, plus I cut it horizontally, B times this. Notice this is simplified to A times 2 vertical strands plus B times 2 horizontal strands. Thus, I can substitute the work I've done for my first part and the work I've done for my second part into my equation that talks about how the Reidemeister II works.

Let's try this. When I do this, and I substitute, I can combine like terms. My very first part becomes the 2 strands crossing like this for my Reidemeister II equals A^2 times a horizontal cut, of these 2 strands, plus ABC times another horizontal cut—I've combined like terms from algebra before—plus AB times a vertical cut plus B^2 times a horizontal cut by combining like terms in the second part.

If I simplify this by looking at all the like terms, you see that you have a bunch of terms which have this horizontal cut and a bunch of them that have this vertical cut. You're combining all of them with the horizontal, you get that these 2 Reidemeister II strand cuts equal $A^2 + ABC + B^2$, the quantity times this horizontal cut plus AB times a vertical cut. Great! I've just done algebraic manipulation. You might want to watch this part again to get a more detailed understanding of what happened with these pictures. But, I've simply performed rules 1, 2, and 3 so far.

If you look at this equation that we have at the end, we notice that we want this to equal 2 vertical strands. Remember what Reidemeister II says. If you have these 2 strands that cross like this, we want it to become 2 vertical strands. I want that. But, in order to make that work, the value in front of the 2 horizontal strands must be a 0 value. It must completely disappear. Moreover, the value in terms of my vertical strands must have value 1 because I want this to equal this. I want Reidemeister II to work. How can I do this?

Let's take a look. I need the second value, AB times the 2 vertical cuts, to actually equal 1, which means A times B has to equal 1, which means solving for B. B has to be 1 divided by A. Another way of saying it is B equals A^{-1}. We also know from the other part that $A^2 + ABC + B^2$ has to equal 0. This was the term in front of the 2 horizontal cuts. When I do a Reidemeister move and if I want the Reidemeister move to not change my polynomial value, my horizontal cut value has to completely disappear because I want the polynomial value for a Reidemeister Move II to become the same regardless of which way I do it.

Thus $A^2 + ABC + B^2$ has to equal 0. But, I just found out what B was. I'm going to plug that in. This becomes $A^2 + C + A^{-2}$equals 0. Now I can solve for C and I get C equals negative the quantity $A^2 + A^{-2}$. Remember how we started off—those 3 rules and based on those 3 rules, we picked values of A, B, and C arbitrarily. We didn't know they were just placeholders and we wanted to know how far we could push those 3 rules. By choosing those random values of ABC for the placeholders in order to make my Reidemeister Move II work, what do I have to do?

I have the relationship of what B has to be A^{-1} and I also have to have the relationship of what C has to be, $-A^2 + A^{-2}$. Thus, this is forced on me if I want Reidemeister Move II to work, which means I have a new set of rules. Let's take a look.

Under this new set of rules, rule number 1 hasn't changed at all because we really didn't use it for the calculation of Reidemeister Move II. But, the rule number 2, when we have this crossing, becomes A times the vertical cut plus A inverse times a horizontal cut. Another way of saying it is if you have a negative slope crossing, we get A inverse times a vertical cut plus A times a horizontal cut. I've just substituted anywhere I saw a B, I now write A inverse. Rule number 3 becomes anytime I see a link with this extra perfect circle, I can throw that circle out, but I need to multiply this term of $-A^2 + A^{-2}$ every time I do it. That's great! All of this work just to try to make Reidemeister Move II happy.

What about Reidemeister Move III? I have those 3 variables. Now, I only have 1 variable left over because the other 2 were already determined by making Reidemeister II work. Let's try Reidemeister Move III. Let's see what happens. Let's take a look.

If we have Reidemeister Move III, I have my crossing with my strand on 1 side and I'm going to push my strand to the other. I need to make this also work. Let's try it. We see that the bracket polynomial of my Reidemeister Move III equals—let's pick that central crossing where we have that positive slope—A times—now I'm going to cut that up vertically—plus A inverse. I'm going to cut it horizontally. Now look at that first picture, the first picture I can do just a Reidemeister Move II without changing the polynomial. I worked extremely hard to make it so. I'm just going to simply move a Reidemeister Move II straight across.

The second picture, I'm basically going to move the strand over. This is just an isotopy move, there's no crossing change. I'm just going to push it to the other side for visual coolness. Let's go back to the first picture again. Now, I have 3 vertical strands. I'm going to take the second and third and take the second and cross it over the third. I'm going to do another Reidemeister Move II, but since we worked so hard, we know that this move comes for free again. We can do Reidemeister Move IIs without changing the polynomial. Thus, we have this new equation of A times this particular picture plus A inverse times this particular picture.

What is this? If we look carefully, this equals just the bracket polynomial of this picture. Why? Because if I take that crossing right there and if I look at it—it's a positive slope—if I cross it vertically and horizontally, I get exactly what I have at this equation. What does this mean? This means that if I have the bracket polynomial of my Reidemeister Move III move setup, I get the Reidemeister Move III working out without conflict.

In other words, Reidemeister Move III comes for free. Unlike Reidemeister Move II, where I had to work, III just comes for free. It's fantastic! I just need to do 1 thing and my life is set. I need to do Reidemeister Move I. If I can show that the bracket polynomial does not change because of Reidemeister Moves I, II, and III, I win. I've shown it works beautifully for Reidemeister Move II. I worked hard to make it so. It works elegantly for Reidemeister Move III; it just falls out of this picture.

What about Reidemeister Move I? Let's take a look. If I have a Reidemeister Move I twist, I want the bracket polynomial of this thing to be a vertical line. What happens? I pick this crossing—it's a positive crossing—and cut it vertically. I cut it horizontally. I get 2 new pictures. The first one has that knot, that elegant unknot. I could just throw it away, but I need to multiply it by negative. $A^2 + A^{-2}$. The second 1 is just a straight line. It's fantastic!

In fact, if you look at this, I have 2 bracket polynomials of the same thing. Thus, I can combine like terms and when I do so, I can simplify it to get $-A^3 - A^{-1} + A^{-1}$. The A^{-1} s cancel out and I end up with $-A^3$ times the bracket of the straight line. Now my stomach starts turning. It doesn't feel good. Why? Because I had the bracket of the twist equals the bracket of the vertical—the Reidemeister Move I—but they're not equal because I have this $-A^3$ in front of it.

In fact, if I did the other Reidemeister Move I twist, if I twisted it the opposite way, you'd get this picture. Notice here I'm doing the exact same kind of algebraic calculations, but at the end, I get $-A^{-3}$ of that bracket. Look at the previous one. In the previous one, when I did the first twist, I get $-A^3$. Now I get $-A^{-3}$. Somehow it works for Reidemeister Moves II and works for Reidemeister Moves III, but it fails for I.

What does a mathematician do? We say, what is it about Reidemeister Move I that's making it fail? Can we somehow measure the Reidemeister Move oneness of this? Then we think and we say, this is exactly what the writhe did. The Reidemeister Move I moves changes the writhe by exactly plus or minus 1. In fact, the writhe was what was frustrating us before because everything worked but Reidemeister Move I. Now we see everything again works but Reidemeister Move I. Maybe we can combine both of these negative properties that Reidemeister Move Is keep failing and make it work beautifully. It turns out that this is what happens.

We define a new polynomial called the X polynomial as follows. We define the X polynomial of the link to equal $-A^3$ raised to the negative, the writhe of the link multiplied by the bracket. In some sense, this funny term of this $-A^3$ raised to the negative writhe is my Reidemeister Move I buffer. What this exactly does is it's supposed to take the failure of the bracket polynomial which could not measure the Reidemeister Move oneness correctly. It compensates with this writhe term, which is also measuring the Reidemeister Move I and together you multiply them and we hope it cancels out perfectly.

Notice the Reidemeister Move II and the Reidemeister Move IIIs are not affected by the writhe or the bracket. We worked hard to show that the writhe didn't make a difference between Reidemeister Move II and III and neither did the bracket polynomial. There's only the Reidemeister Move Is where we got a stumbling block. This X polynomial is a knot invariant.

Let's take a look. Look what happens if I take the X polynomial of a Reidemeister Move I twist. I get $-A^3$ times negative the writhe of this twist times the bracket polynomial of this twist. The writhe of the twist is exactly the writhe without a twist with a plus 1. You can calculate this on your own. I can substitute negative instead of a writhe of a twist, I get negative the quantity writhe of a vertical line plus 1 times the twist of the bracket. Great! But, if you notice the twist of the bracket, I can actually untwist the bracket as long as I change my crossing values. By doing so, I get $-A^3$ which is what we had computed earlier when we take care of this twist.

What do I have? I can combine like terms again and I get $-A^3$ times negative the writhe of that vertical minus 1 plus 1 times the bracket of that vertical strand. But, I can simplify the top power exponential power. This is $-A^3$ the writhe, negative writhe of that vertical, times the bracket of the vertical. This is exactly the definition of the X polynomial for a vertical line. In other words, if I have the X polynomial of a twist, I actually end up with it being the same thing as the X polynomial of that vertical strand. The twisting does not change X. What the bracket gives me to make it fail is exactly what the writhe provides to make me succeed.

We don't need to worry about Reidemeister Moves II and III because we know they work for both the writhe and the bracket. It's only I that it fails for and this shows it succeed beautifully. If you had taken an opposite twist, you would end up with exactly the same value—that it does not matter at the end of the day.

Let's close with some comments. We have learned a lot of material here, combining and actually building this invariant from scratch. It was hard work and I encourage you to look through this lecture again to see the small details that might have been pushed over. If you love algebra, this is a great lecture to go back and film the details. If you don't love algebra, you can assume that what I'm saying is right. We have constructed this powerful knot invariant from a projection of a knot. It turns out that it does not depend on the projection itself. It's an honest to goodness invariant.

Officially, technically, to get the Jones polynomial, what you do is take the X polynomial we've been talking about and every time you see an A, you substitute 1 divided by T to the ¼ power to the fourth root of T. This is a very awkward substitution, but if you do the substitution, you will get the Jones polynomial on the nose. But, for us, it doesn't matter. It's basically the same thing with a little change of variables. Instead of calling it T to some power, we're going to call it A to some power. This A is far more intuitive for us because we actually saw how it came about. We're going to ignore this technicality from now on.

How powerful is this polynomial? A big, open question has been—which I asked several lectures ago—can we tell any knot from the unknot? Any time we use the Jones polynomial, we get the value of the unknot as 1. But, any time we use the Jones polynomial of any other knot, which turns out to be not the unknot, the Jones polynomial is not 1. It seems that the Jones polynomial is able to tell every knot apart from the unknot. This is still an open question. This beautiful invariant that we've created turns out to be extremely powerful.

Next lecture we're going to actually use this to perform some calculations to get our hands a little bit dirty. See you then. Stay tuned.

Symmetries of Molecules

Lecture 7

A knot or a link is called amphicheiral if it can be made or deformed into its mirror image. What we're interested in is examining the mirror images of knots since they might give different chemical properties for these topological stereoisomers.

A broad topic in the study of shapes is the idea of symmetry and the question of whether the mirror images of 2 objects are equivalent. This lecture shows how this question relates to work on molecular compounds and topological stereoisomers in chemistry.

We begin with a review of calculations of the X-polynomial. For example, we compute the bracket polynomial of a double twist, the unknot, the Hoff link, and the trefoil, as well as the X-polynomial of the double twist and the trefoil. Note that any time we have a complicated knot, if we know how it works previously in a simpler version, we can use that value to compute the X-polynomial of the more complicated knot.

The X-polynomial has another stunning feature that relates to additions of knots. In a previous lecture, we asked the question: How are the crossing number and the unknotting number related to knot addition? The beautiful feature of the X-polynomial is this: The X-polynomial of knot 1 + X-polynomial of knot 2 = the X-polynomial of knot 1 × the X-polynomial of knot 2. Once we understand how prime knots work—the basic building blocks of knots—we can get polynomials for composite knots by this simple procedure. If we have a complicated composite knot, we just break it up into its prime pieces, compute each one separately, and multiply the answers together.

Before we continue, let's look at a pair of molecules, each with an identical number of atoms and identical atomic bonds but different placement in space. Such pairs are called topological stereoisomers. From a chemistry point of view, these 2 seemingly identical objects might have different properties. Chemists are interested in topological stereoisomers because they provide a means to create entirely new substances. In the study of knots, a knot or a link is called **amphicheiral** if it can be made or deformed into its mirror image. The mirror images of knots might give different chemical properties for these topological stereoisomers.

We see that Reidemeister moves can yield a mirror image of the figure 8 knot but not the trefoil. Notice that a projection of a knot and its mirror image are identical. They look the same, except that every positive crossing is replaced by a negative crossing. Thus, we need a tool that can tell whether each crossing is negative or positive. It turns out that the X-polynomial is beautifully designed to help do that.

The way multiplication works of polynomials somehow captures exactly how knots are put together. ... The way these knots are glued and the way this new knot is formed is exactly captured by polynomial multiplication. This is gorgeous! It is a stunning and elegant result.

For any knot and its mirror image to be amphicheiral, the X-polynomial must be a palindrome. For example, the X-polynomial for the figure 8 knot is: $a^8 + a^4 - 3 + a^{-4} + a^{-8}$. If we take a mirror image of this knot, then all the a's become a inverses and all the a inverses become a's; we get the same polynomial because the figure 8 and its mirror image are the same.

Earlier, we calculated the value of the X-polynomial of the trefoil to be $-a^{16} + a^{12} + a^4$. By switching a and a inverse, we get $-a^{-16} + a^{-12} + a^{-4}$, which is the X-polynomial for the mirror image of the trefoil. These 2 polynomials aren't the same, which means that the mirror images of the trefoils are fundamentally different. Only the X-polynomial gives us the power to find this result. ■

Important Term

amphicheiral: An object is amphicheiral if it can be made into its mirror image.

Suggested Reading

Adams, *The Knot Book.*

Questions to Consider

1. What properties or traits do we use to tell an object apart from its mirror image? Which objects in our daily lives are equivalent to their mirror images?

2. Show that $X(\text{trefoil} + \text{trefoil}) = X(\text{trefoil})X(\text{trefoil})$

Symmetries of Molecules

Lecture 7—Transcript

Welcome back and thanks for joining me again. A broad question in the study of shapes is the idea of symmetry. Consider humans or most animals. We are extremely symmetric beings where there are 2 parts to us: The right side and the left side. Every one of us has a pair of hands—the right side and the left side—which are practically identical; 1 is a mirror image of the other one.

What scientists are interested in are objects which are equivalent to their mirror image. Is my right hand equivalent to my left hand? No matter what I do to take my right hand and make it into my left hand, I can't succeed, because my thumbs—although they fit together perfectly—if I match them up the way they should in terms of the way the skin of the inside and the outside is, it doesn't work. They're mirror images, but they're not equivalent.

As we try to distinguish shapes throughout these collections of lectures, we notice that this is one of the hardest things to do. Why is that? It's because things which are mirror images are practically identical. Every property that 1 hand has, the other hand also has. Does it have a thumb? Yes, it does. Do they bend the same way? Yes, they do. Does it have a palm? Yes it does. Any property you can think of, they both have. How are we trying to succeed in this particular venture of telling 2 objects whose mirror images could be different or could be the same? This is the frustration that scientists are going through. This is where math is going to step in to help us.

We show in today's lecture how this relates to work on molecular compounds and topological stereoisomers in chemistry. And we want to use the X polynomial that we spent the previous lecture talking about, that we spent the previous lecture constructing from scratch to help us with this.

We'll now review our work on the X polynomial by doing some calculations. Last time we built it up, but we really didn't get a chance to use it. The first part of the lecture is going to be based on actually using the X polynomial. Remember the X polynomial and the Jones polynomial are basically identical except for just a notation of calling 1 A and calling the other one a little change of variable using T.

Let's begin by computing the X polynomial for the unknot. We're not going to look at the simple unknot, which looks like a circle. We're going to look at the X polynomial of an unknot with a few twists in it. Notice here that the X polynomial of the unknot is what we are trying to find. We already know the answer to this puzzle. We know that the X polynomial is a knot invariant, which means any projection of this particular knot will be the same X polynomial for us.

For this particular knot, which happens to be the unknot, we already know the X polynomial is 1, so let's find out if we can build this X polynomial up from this twisted projection and see what we get. Let's compute this.

Compute the X polynomial; we know we need this to equal $-A^3$ to the negative writhe of this particular projection times the bracket of this projection. We need 2 pieces so compute the writhe and compute the bracket. The writhe of this particular projection, if you just pick an orientation, is going to give you a value of 2. The bracket of this polynomial—which we're going to show in a little bit—becomes A raised to the sixth power. Let's combine these 2 things and see what we can get for the X polynomial.

The X polynomial of this twisted unknot equals $-A^3$ the quantity to the negative 2 times A^6. Thus this can be simplified as A^{-6} times A^6 but this equals 1. We get what we want, that it should be 1 as we have expected it to be. One of the steps that I skipped here was computing the bracket for this twisted unknot. Let's actually look at the zoomed-in version of what the bracket of this unknot becomes. Let's take a look at this picture.

Remember the polynomial relationship we get by rule number 2. If I take the first crossing, which is a positive crossing, a positive slope crossing, I can cut it up vertically first, and I can cut it up horizontally, and I get equals A times this figure plus A inverse times this figure. The first figure, though, has that circle in it, that pure, perfect circle. I can throw it away and I need to multiply it by negative the quantity A^2 plus A^{-2}. I do so in the first one.

The second 1 is just a simple twisted unknot. Notice they both have the same twisted picture for both of them for the bracket. I can combine thus the like terms, and I end up with this unknot with this double twist equals $-A^3$ with a single twist.

We can do the same procedure again: I take the crossing that I see—it's a positive crossing—I cut it vertically, I cut it horizontally, and I get A times the quantity of 2 circles plus A inverse times the quantity of 1 circle, the whole term multiplied by $-A^3$, which equals simplification of one more. I can throw out the inside circle with again a negative, the quantity $A^2 + A^{-2}$ and now I can simplify everything because they all have the term of the bracket of the perfect circle.

If I simplify everything, I get this value equals $-A^3 \times$ another $-A^3$. We keep picking up these $-A^3$ for these twists, which the writhe is building for us, $-A^3 \times -A^3$ times the value of the perfect circle, which we know is 1 by rule number 1. This equals A^6, which is great! We got the value we used the last time in the previous slide to talk about the value of the X polynomial of the double-twisted unknot.

What about a link? We've been talking about this unknot before in terms of this double twist, but what if we throw something more complicated like the Hopf link; can this handle it? It absolutely can! Let's take a look.

Look at the bracket polynomial for the Hopf link. Notice that it has 2 crossings. Let's pick 1 of these 2 crossings to work with. I'm going to pick this crossing, which has a negative slope, which means my first cut is a horizontal cut and my second one is a vertical one. Since this polynomial is only based on the projection of this knot, since we are talking about the bracket polynomial not the X polynomial, the bracket 1 fixes our particular projection. As we continue with this calculation, we see that this equal A $\times$ the quantity A $\times$ the bracket of 2 circles plus A inverse $\times$ the bracket of 1 circle + the second terms—again we can resolve one of these crossings by choosing the crossing and cutting it horizontally and vertically and I get its A inverse $\times$ the quantity A $\times$ the bracket of 1 circle plus A inverse $\times$ the bracket of 2 circles.

We can group all the like terms together again using algebra. Grouping the like terms which have 2 circles and those that have 1 circle, we get the quantity that this equals: A^2 plus A^{-2} the quantity × the bracket of 2 circles + 2 × the bracket of 1 circle; but if we have 2 circles I could just pop 1 out, throw it away, as long as I multiply by the corresponding value of negative the quantity $A^2 + A^{-2}$. This is rule number 3. I get only 1 circle at the end of the day, group all my like terms together, and I get the value negative the quantity $A^4 + A^{-4}$.

We have been able to compute the bracket polynomial of a double twist, of the unknot, the bracket polynomial of the Hopf link, and we even computed the X polynomial of the double twist. Let's be a little bit more ambitious; it seems like we haven't done anything that exciting. Let's compute the X polynomial of a knot that we know and love, the trefoil, the simplest shoestring knot we started this lecture with.

If you look at the bracket polynomial of the trefoil, notice what we get. Consider this picture: The bracket polynomial of the trefoil equals its $-A^3$ to the quantity negative the writhe of that particular projection × the bracket of the projection. This is the value of how we compute the X polynomial. What is the writhe of the trefoil? Of this particular projection of the trefoil, we get that the writhe is −3. We can choose the orientation we want and use the classic value of how we measure writhe. If it's a positive crossing it's +1 and negative crossing it's −1 and we add it all up.

What is the bracket polynomial of the trefoil? Remember we need the writhe and the bracket together to get this answer for the X polynomial. The bracket polynomial of the trefoil is you have to pick a crossing. I choose the very bottom crossing, which is a positive crossing. I'm going to cut it vertically, and then I'm going to cut it horizontally to get this relationship of these 3 diagrams.

Look at that diagram of the 1 next to A. Notice that this is a double twisted unknot. We already computed the bracket for this; this is the first thing we did—this was A^6. Look at the second bracket of that figure; this is the Hopf link, and we just computed this. This is the value negative the quantity $A^4 + A^{-4}$. We just plugged those values in because we just did the work to do this, and we get the answer to be $A^7 - A^3 - A^{-5}$; this is the bracket polynomial. If we take the bracket and plug it into our original formula to get the X, we need to compensate the writhe. By doing so we get, the X polynomial of the trefoil becomes $-A^9 \times$ the quantity $A^7 - A^3 - A^{-5}$ which equals if we simplify it $-A^{16} + A^{12} + A^4$.

The X polynomial has another stunning feature. Notice that we have computed the X polynomial of the trefoil and it took a lot of work. As we computed this we note that we actually started using previous computations in this computation of the trefoil itself. This is also going to be the case for X polynomial computations in the future. Any time you have a complicated knot, if you know how it works previously in a simpler version of it, you can use that value in computing the X polynomial of your knot. The stunning feature, this beautiful result of the X polynomial, does not lie in how it computes things, but how the X polynomial relates to additions of knots.

Remember how we talked about knot addition and we wanted to know this big unsolved question, but how the crossing number is related to knot addition and how the unknotting number is related to knot addition? Here's the way the X polynomial relates to additions of knots. Listen and cry when I share this theorem with tears of joy. The X polynomial of knot 1+ knot 2 = the X polynomial of knot 1 × the X polynomial of knot 2—what an amazing result.

You might be wondering why this is exciting. Think about how polynomial multiplication works; it's not that simple. If I have a polynomial of 2 terms, and I multiply it with another polynomial of 3 terms, how do I do this? I take my first term and multiply it with the first one, first with the second, first with the third, and then I take the second with the first, the second with the second, and the second with the third. Then I add up all these terms together. This is the way polynomial multiplication works.

What does this result that I've just shown you say? It says if you take the first knot and find its X polynomial, then if you take the second knot and find its X polynomial, then if I put these 2 knots together along the way we defined addition, this new knot I have has the X polynomial of these 2 polynomials multiplied together. The way multiplication works of polynomials somehow captures exactly how knots are put together. This is stunning. The way these knots are glued and the way this new knot is form is exactly captured by polynomial multiplication. This is gorgeous! It is a stunning and elegant result.

What we see is that once we understand how prime knots work—the basic building blocks of knots—we can get polynomials for composite knots by this simple procedure. Somebody gives you a complicated composite knot, you just break it up into its prime pieces, compute each 1 separately and just multiply the answer together; it's fantastic.

We see how powerful the X polynomial is, but we want to use it to understand issues with symmetry. Remember that's how we were motivating this talk, by the right and the left hand. What does symmetry have to do with science and nature? A chain of atoms with the same bonds in exactly the same sequence may turn out to form different molecules.

Let me give you an example: Consider this example of 2 molecules, as you see in this figure, made up of twisted ladders with 4 rungs: 1 with a left twist and another with a right twist. Notice that the number of atoms here in this twisted ladder are both identical for both of them, and the way the atoms are glued together are identical. The kinds of bonds are also identical; but one of them—the top one—has a twist with one of the rungs going on top of it, and the bottom one has a twist with one of these rungs—or one of these bonds—going the other way. In other words, they both have the same molecular structure with atoms and bonds in the same order; however they're embedded, or placed, in space differently.

Such a pair of molecules are called a pair of topological stereoisomers. From a chemistry point of view, these 2 seemingly identical objects might turn out to have different properties. One of these might have the property of a liquid—like water—and another one might have a property like oil. The shape of the molecule determines structure; form and function are related.

Chemists are very interested in topological stereoisomers because they provide a means to obtain substances possibly never seen before. If you take the same collection of atoms and the same collection of bonds and put them together in different ways in terms of the way they show up in space, you might end up with different molecular structures. How do we get stereoisomers? It turns out we make them.

The process of creating and synthesizing these molecules have a rich history. We have seen that DNA is a molecule made up of millions and millions of atoms. This knotting of DNA structure is easy because you have so much to work with. The question is: Can we do this for smaller molecules? Any knot made from chains of identical atoms and bond is topological stereoisomer to any other knot. Why is this? It's because knots at the end of the day are just circles. If you're going to make a knot, it's just a collection of atoms with bonds between it that form a loop.

Every knot is going to be a topological isomer to any other knot that you can make. Thus, a scientific race began to synthesize knots from atoms, because if you can make different knots from the same pieces of the puzzle, you might end up with different molecular structure stereoisomers that might give you different properties, like water and oil.

A catenane is a set of linked molecular rings; it's not a knot, but it's a link in some sense. The first successful synthesis was created by Wasserman in 1960. He used at least 20 atoms to do this. Thus, linked molecules were created and this is easy to do because they're bigger objects that you can use. What about knots? What about just 1 piece that you can make out of this? For knots, unless a large amount of atoms are present the molecular strand is too inflexible to tie in a knot.

Let me explain to you what that means. If you have atoms with molecular bonds between them, since the bonds cannot be closed up too much because the bonding angle is going to make it open up more--because of this inflexibility—then I don't have the freedom to make sharp bends with my atom bond chain to close up to a knot. Since the bond angle forces me to stay open as much as possible, I need to have a lot of atoms with the bond angles pretty open in order to close up to actually form a knot.

However, Christine Dietrich-Buchecker and Jean-Pierre Sauvage succeeded in creating the first synthesis of a knotted molecule, the trefoil, in 1988. The race was finally won. In 1990, just 2 years later, Qun Yi Zheng and

David Walba used techniques of creating specially twisted ladders—some techniques that we saw earlier in terms of how we created these 2 separate rings—to create more knotted molecules. Currently this is still a hot area of research and not much has been known, but at least the main barrier has been broken.

Let's go back to our motivating question concerning mirror images of objects. A knot or a link is called amphicheiral if it can be made or deformed into its mirror image. What we're interested in is examining the mirror images of knots since they might give different chemical properties for these topological stereoisomers. What about the figure 8 knot? Let's look at this picture.

Is the figure 8 knot the same as its mirror image? Can we take the figure 8 knot, move it around, and get the same figure 8 knot with every positive crossing now a negative 1, 1 with positive slope now a negative slope? Is this possible using Reidemeister moves? Let's take a look at this demo.

Here you see a figure 8 knot, and as you're looking at it I want you to look at the crossing information. Notice this crossing is going under this strand, and this crossing is going above this strand, and all the crossings become alternating, going under and above and under and above. Using simple moves here, we are able to get the figure 8 knot just like we were before; it looks identical to it. Now, though, my crossing information has changed. Here you see the crossing is now going over it, and this strand is now going under this. It's exactly what we had before, but using a really simple set of moves, we were able to get the figure 8 knot, its mirror image using simple Reidemeister moves—just moves in space.

What can we say about the trefoil knot? In other words, if you look at this figure, is the trefoil the same as its mirror image? We saw for the figure 8 that the figure 8 and its mirror image are the same. You can take the figure 8 knot, move it around in space, and get its mirror image without cutting and gluing—they're the same knot. Is the trefoil, though, the same as its mirror image?

If I take a trefoil in my hand and move it around as many times as I want to, it turns out I cannot--unlike the figure 8 knot—make it into its mirror image. No set of obvious moves exist that I can do this for, but this doesn't mean there are some complicated collection of moves that it won't work for. How can I tell whether the trefoil and its mirror image are the same or different? For the figure 8 I was lucky; it was an elegant set of moves. Again it seems like we're trying to measure this quantity of amphicheirality—mirror imageness.

Consider this in terms of projections. Notice that a projection of a knot and its mirror image are identical. They look the same, except that every positive crossing is replaced by a negative crossing. Thus, we need a machine, we need a weapon, we need a tool that is sensitive enough to this crossing information that it doesn't worry about the bigger picture, but somehow can capture the crossing information at each one of these crossings and tell whether it's one kind of crossing or another one.

It turns out that the X polynomial is beautifully designed to help us do this. Let's take a look at this example and see what I mean. Consider the bracket polynomial of this particular positive crossing. What do we know? The bracket of this positive crossing equals A × a vertical split + A inverse × a horizontal split. This is exactly rule number 2. Let's rotate my head 90 degrees. In other words, let's look at this exact same picture, but now instead of a positive crossing, let's look at a negative crossing.

Here, the bracket polynomial of this negative crossing equals A × the bracket of the horizontal split + A inverse × the vertical split. I'm going to take these last 2 terms and I'm just going to switch their places. Instead of A × something + A inverse, I'm just going to switch their places, and I get that the bracket of a negative crossing equals A inverse × a vertical split + A × a horizontal split. Compare this last equation to the first one we came up with, with a positive crossing. Notice that every time I see a positive crossing I get A vertical A inverse horizontal. However, if I have negative crossing I got A inverse vertical A horizontal.

What does this mean? This means that changing one crossing with another one, changing a positive crossing to a negative crossing in the same bracket polynomial, switches A and A inverse around. Knot K and its mirror image K Star have the same X polynomial with A and A inverse switched. In other words, given any knot, you can find its bracket polynomial and take its mirror image, and instead of actually working hard to find out its bracket polynomial, we just look at the original bracket and any time you see an A you put an A inverse, and any time you see an A inverse you put an A. This is what rule number 2 says for us.

What happens if this knot that I'm looking at is amphicheiral? Then the knot is its mirror image; they're the same thing. That means since the bracket polynomial, which gives us the X polynomial—since the X polynomial is a knot invariant, this means that the knot and its mirror image since they're the same--must have the same X polynomial. We just got done saying, though, that any knot and its mirror image must have X polynomials with the A and A inverses switched. Therefore, if K and K Star have the same bracket polynomial, where switching A and A inverse does not change the polynomial, think about that.

You must have—in order for K and K Star to have the same knot, in other words for a knot to be amphicheiral—the X polynomial must be a palindrome. What is a palindrome? A palindrome is something that is the same read in 1 direction as in the opposite direction. For example, the phrase "never odd or even" is a palindrome. I could read it normally, "never odd or even" or I can read it backwards, and I still say "never odd or even."

For example, the figure 8 knot has an X polynomial as $A^8 + A^4 - 3 + A^{-4} + A^{-8}$. Look what happens if I change all my As to A inverses. If I take a mirror image of this polynomial, of this knot, then all my As become A inverses, all my A inverses become As, and I get the same polynomial because the figure 8 and its mirror image are the same. We just showed this.

What are the consequences of this result for other things in mathematics? What about the trefoil? Earlier today we calculated the value of the X polynomial of the trefoil to be $-A^{16} + A^{12} + A^4$. By switching A and A inverse we get a polynomial $-A^{-16} + A^{-12} + A^{-4}$, which is the X polynomial for the mirror image of the trefoil. These 2 polynomials aren't the same, though. Since 1 polynomial and the other polynomial aren't the same, that means that the knots that we're getting based on this polynomial must not be the same; which means that there are 2 kinds of trefoils—there's the right-handed trefoil and the left-handed trefoil. The mirror images of the trefoils are fundamentally different.

There's only one of the figure 8 knot, but for the trefoils, there are 2; what a result! We've always thought that there's only 1 trefoil. I kept talking about THE trefoil. It turns out that there are 2 trefoils. How do you think we could do this without using the X polynomial? Tricolorability, they would look identical. Crossing number, they would also look the same. Unknotting number, all of these things aren't sensitive enough to measure crossing switches of taking all my positive and making it negative, and all my negative and making it positive.

This is the power of the X polynomial; this is what is used in chemistry—ideas like this that push us to the forefront of not just mathematics but chemistry to better understand symmetry for molecules, knowing that if you construct the figure 8 not using chemistry, that's fantastic. You don't need to worry about constructing its mirror image. If you construct the trefoil knot in chemistry then—if you construct its mirror image—you actually get something fundamentally different. It's actually worth it to pursue both kinds of constructions.

Although amphicheiral implies palindromic—in other words, if you're not as amphicheiral—then the X polynomial must be palindromic, the converse of this is not true. Consider this picture: Here we have an example of a knot—it's the 42 knot with 9 crossings. One can show its X polynomial is a palindrome—if you actually compute the X polynomial of this you get a palindrome. Using a powerful notion in mathematics called a signature invariant—far beyond what this lecture can do—1 can show that this knot is not amphicheiral. It might have a palindromic X polynomial, but that doesn't mean amphicheirality. It only means the 1 direction, not the other. In other words if you know your knots are amphicheiral, then your knots they have to palindromic. However, if they're palindromic in terms of their X polynomial, we don't know anything about amphicheirality.

We close this lecture with a big open question, useful for biology, chemistry, and mathematics. Can you find the complete invariant that measures amphicheirality? The X polynomial helps us to go in this direction, but it isn't strong enough to tell everything apart, like this knot we just talked about.

In closing we have learned to use the X polynomial to perform computations. It is difficult and it's cumbersome to go through these calculations in terms of talking about writhe, and in terms of talking about the bracket polynomial, but the more we start doing these computations, the more useful they become in computing other more-complicated knots. We've also seen that topological stereoisomers are extremely important in chemistry, able to create different possible objects maybe with properties of water or oil, and mirror images of objects provide us with a way of creating them. We have applied the power of the X polynomial to partially distinguish knots from their mirror images.

In the next lecture we leave the world of chemistry and move into the world of biology with a mathematical study of mutations. Stay tuned.

The Messy Business of Tangles and Mutations

Lecture 8

For Conway, there are 3 main types of tangles for us to focus on. ... There's the infinity tangle. ... There's the 0 tangle, ... and then based on these 2 tangles, Conway said you can come up with the *n* tangle, where *n* represents any integer that you want.

Your cells, as they're dividing and multiplying, need to copy DNA information located in one cell into another. But as we know, DNA is a tangled mess. How can this structure be copied without too many errors? The answer is that an enzyme called topoisomerase alters the topology of the DNA. It straightens the DNA at a small local area to make the copying easier, then cuts, twists, changes crossings, and so on.

Error is sometimes introduced in the copying process, and when it is, mutations occur. In this lecture, we will look at the mathematical version of mutations. The mathematical notion of tangles, developed by **John Horton Conway** in the late 1960s, is one means of understanding mutations. A tangle is defined as a part of a projection of a knot, or link, around the circle, crossing it exactly 4 times.

The Reidemeister moves can be used with tangles under one condition: Tangles can be moved around only inside the circle. We say that 2 tangles are equivalent if one can be made into the other by Reidemeister moves within the circle. We can connect any tangle and make it into a knot or a link by connecting together the 2 northern and the 2 southern strands.

Conway's notation for knots starts with the simplest tangles and builds to more complicated ones. For Conway, there are 3 main types of tangles to focus on: the infinity tangle, the 0 tangle, and the n tangle, where n represents any integer. For example, a 2 tangle is simply a tangle that has 2 twists in it. To distinguish between a $-n$ tangle and a $+n$ tangle, we look at the overcrossing. If it has a positive slope, the tangle has a positive value; if it has a negative slope, the tangle has a negative value. This is the building block that Conway uses to construct rational tangles. Such tangles can be constructed using rotations and reflections or with a method based on the number of integers in the notation. However, the notation itself is not a complete tangle invariant.

Conway's theorem relating to the equivalence of rational tangles states: Two rational tangles are equivalent if and only if their continued fraction values (derived from the notation) are the same. Remarkably, the continued fraction value embodies the shape of the tangle itself. Moreover, knots or links formed from these tangles will also be identical.

... if you're given 2 tangles with the Conway notation, if you can find the continued fraction values and they turn out to be the same, this means that the tangles—the pictures themselves—must be the same.

Tangles lend themselves beautifully to the arithmetic operations of addition and multiplication. Any tangle formed by addition and multiplication is called an algebraic tangle. Just as numbers do, tangles have an additive identity, the 0 tangle, but the multiplicative identity and the associative property fail for tangles.

We can perform 3 possible mutations of a tangle in a knot. These mutations are a mathematical approach to capturing what happens in DNA structures. Unfortunately, mutation operations are extremely hard to distinguish. Greater tools are needed to crack this problem in both mathematics and nature. ■

Name to Know

Conway, John H. (1937–): Conway is a professor at Princeton and a prolific mathematician whose works encompass geometry, group theory, number theory, and algebra. In particular, he is known for his Conway notation for knots and links.

Suggested Reading

Adams, *The Knot Book*.

Questions to Consider

1. Construct some mutations and show that the Jones polynomial remains the same for both mutants.

2. What does mutation do to the Conway notation of a knot?

The Messy Business of Tangles and Mutations

Lecture 8—Transcript

Welcome back and thanks for joining me again. The DNA structure is used to perform a vast number of biological functions. It is basically the blueprint of life. One of the things that DNA does, millions and millions, of times is to be copied. Your cells, as they're dividing and multiplying and growing, need to copy the DNA information located in one cell into another one. As we have talked about before, the DNA is a tangled mess. It's packed into the nucleus of your molecule. How are you going to copy this tangled up structure without causing too many errors? It's made up of millions of atoms put together.

An enzyme called topoisomerase alters the topology of the DNA. One of the first things it does is it comes in, it stretches the DNA at a small local area to be straight, so the copying is much easier at that part. Then it moves into the DNA structure itself. It sometimes cuts, adds, twists, changes crossings, or does far more complicated operations based on the kind of enzyme that you're dealing with.

We had mentioned this earlier in a previous lecture when we discussed the unknotting number. Remember how we wanted to cut the knot to change the crossing from overcrossings to undercrossings and vice versa? The particular type of action of an enzyme is of high importance, and this is a site-specific recombination. The enzyme aligns the 2 strands up, cuts the strands open, then reattaches the ends in a different possible way. That's why it's called site-specific, since it's a local operation. It only does it at 1 part of the entire complicated structure.

When things start going wrong, mutations occur. In other words, the way the copying should be done, error is introduced in this process, and any kind of error is what a mutation is. Of course our body has a lot of ways of correcting the mutation because errors happen all the time.

Today we want to introduce the mathematical version of what mutation would be, and it is motivated by this idea of site-specific recombination, focusing at this one particular part. The introduction of the mathematical notion of tangles is one means of understanding such enzyme operations. This notion of tangles was introduced by John Horton Conway in the late 1960s.

I just want to pause and just share with you a minute about this professor. Conway is a professor at Princeton and a genius at discovering underlying patterns to complicated problems. He's the only professor of Princeton that I know who has 2 offices that are next to each other: One for him and 1 for all of his toys. The office is filled with a collection of lots of different puzzles, trinkets, and toys that he works with, and that is actually motivating his work.

When it comes to tangles, Conway was motivated to come up with a notation for knots. How can I describe a knot in an elegant manner? I'll rephrase that question. We have names for special kinds of knots and links—we call some the trefoil, some the figure 8—but then we start running out of ideas. We say 5, 1; 5, 2; 6, 3, based on the number of crossings, the crossing number of the knot itself. Is there a nice way to describe a knot that you've just drawn on a napkin to a friend over the phone? Can you call them up and describe them in an elegant way, other than saying, "Well, what about going under, no I mean over and then move it around?" It's kind of an awkward way of doing it. This is the motivation for Conway's work on tangles. It turns out this is also useful—as we're going to find out—to describing the mathematical way of thinking about mutation. We begin with a definition. What is a tangle? A tangle is a part of a projection of a knot, or link, around the circle, crossing it exactly 4 times.

Let's take a look at this example here: Notice here I don't care about the rest of the knot or the link that's located around this tangle. I only care about that circle that's surrounding this tangle. This is exactly the issue we had with the Reidemeister moves where we locally focus in on 1 part. For Conway, this part is based on 4 special strands that come in, and these 4 strands have corners. We have the northwest corner, the northeast corner, the southwest corner, and the southeast corner.

So these 4 corners can come in as 4 strands and any complicated operation can happen inside, and they leave in these other 2 strands. One of the properties we can do with tangles is we can move them around using the Reidemeister moves that we did with knots except for 1 condition. We can move them around only inside the circle. We say that 2 tangles are equivalent if one can be made into the other by Reidemeister moves within the circle.

Take a look at this example. Notice here I have this tangle. It might look a little complicated, but I'm performing Reidemeister moves within the circle to convert this tangle into this other one. Notice I did the Reidemeister move I twist, and then I can just stretch the strings out by doing a couple of Reidemeister II moves and make it into this tangle over here. In this way these 3 tangles I've drawn are equivalent because I've done all my moves within the circle.

We can connect any tangle and make it into a knot or a link by connecting together the 2 northern and the 2 southern strands. If I connect the 2 northern strands, the northeast and the northwest together and the southeast and the southwest together, then notice I have a closed-up system. Sometimes you might end up with links or sometimes you might end up with 1 omplicated knot.

Conway's notation is based on starting with the simplest tangles and building and constructing to get more complicated ones, just the way we would build numbers. We start with the simplest numbers, and we can get more complicated ones by adding and multiplying numbers. For Conway, there are 3 main types of tangles for us to focus on. Let's take a look.

There's the infinity tangle, and the infinity tangle has 2 vertical lines coming down that connect the northeast to the southeast and the northwest to the southwest. There's the 0 tangle, which is these 2 horizontal lines, and then based on these 2 tangles, Conway said you can come up with the N tangle where N represents any integer that you want. For example a 2 tangle is simply a tangle which has 2 twists in it. You take the 0 tangle that we have and you just make 2 twists; but at the same time you could also have the negative 2 tangle, which is 2 twists the other way.

How are we supposed to know which the −2 tangle is and which the 2 is. There's a simple example that we can think about—which is based on slope—just like we did last time when we talked about the Jones polynomial. Each time you twist you look at the overcrossing, and if the overcrossing has a positive slope, that means you're going to get a plus in your value for the tangle. If your overcrossing has a negative slope, you're going to get a minus. This example right here will give you a +2 twist because it has 2 positive slopes and this tangle we'll say is a −3 tangle. It has 3 negative slopes.

This is the building block that Conway uses to construct any tangle he wants. Here's the way we do it. We start with any N tangle that we have—that you want to begin with—and then we perform rotations and reflections. Let me give you this particular example. Let's say we want to build a tangle called the 2 3 tangle. First we start with a 2 tangle. Then we take the 2 tangle, we rotate it 90 degrees, and then we reflect it. When we reflect it, I don't mean rotate it in space, I actually mean reflect the tangle—take its mirror image. All my crossing information changes from overcrossings to undercrossings, and when I do this I get this following picture: I go from my 2 tangle, I rotate, I reflect, and the moment I have this new answer, then I take my right 2 points—the northeast and the southeast—and then I perform my 3 twist. I start with my 2, rotate, reflect, perform my 3 twist and I get this new 2 3 tangle. We can keep going with this. Let's take a look.

With this I have my 2 3 tangle; what if I want to make it into a 2 3 −1 tangle? I already have a 2 3 tangle. I can take my 2 3 tangle, rotate this entire 2 3 tangle, reflect the entire 2 3 tangle, and now, when I have my new tangle—my new rotated and reflected tangle—I can look at my right 2 points and again do a −1 twist. And now I have 2 3 −1.

In this pattern we can get any tangle we want along Conway's construction, and these tangles that we get based on this work are called rational tangles. Why are they called rational tangles? It's because it is based on this simple sequence of whole integers. We take 2s, 3s, −1s, and get this example; and we can do far more complicated situations if we keep continuing this pattern.

There's an alternate construction of taking and constructing these rational tangles based on the number of integers in our notation. Instead of doing a fancy rotate and twist and reflect, there's an elegant way that Conway has come up with in another way of doing this. Let's take a look.

What we do is, we look at the number of integers in our tangle notation. If our tangle notation has an odd number of integers we start with a 0 tangle, for example the 2 3 −1 has 3 integers—2, 3, and −1, there are 3 of them—and so we start with a 0 tangle and first we twist right—so I twist 2 to the right—and then I twist 3 in the bottom, and then I twist −1 to the right, and I keep continuing the sequence as long as I want to until my number sequence is exhausted. So from my 2 3 −1, I would twist 2, 3, and then −1 and then I'll have my 2 3 −1 tangle. There's no rotation and there's no reflection involved anymore.

What happens if I have an even number of integers? For example what if I have 1 −1 2 1? Here I have 4 things. With an even number of integers I start with the infinity tangle and now I twist bottom. I twist 1, then I twist right −1, bottom 2, right 1, and I keep continuing this bottom right, bottom right sequence. If I had odd numbers, I would do right bottom, right bottom, and if I have even I do bottom right, bottom right. There's no rotation needed and there's no reflection needed.

So here's another shortcut of getting what you want. This is again Conway's creation. What can we say about equivalence of tangles? We already know that 2 tangles are equivalent if you can use Reidemeister moves to go from one to the other one that stays within this circle of power. This circle defines what that tangle is. Is it possible that the notation itself is enough to tell 2 tangles apart? In other words, if you come up with a tangle and you have a notation 2 3 −1 and I have another tangle notation, let's say I have 1 −1 2 1, since our notations are different—our integer sequences are different—does that guarantee our tangles have to be different? Unfortunately it's not that case.

The notation is not a complete tangle invariant. Let's take a look. Consider this example: Here I have a 2 1 1 tangle and I have a −2 2 1 tangle. Notations that are very different. Let's look at the very first tangle here. The 2 1 1 tangle has been constructed this way—remember I take the 0 tangle, twist 2, twist 1, twist 1—twist 2 on the right, 1 on the bottom, 1 on the right—−2 2 1,

I twist 2 on the right negatively, I twist 2 on the bottom, and I twist 1, and I end up with these 2 tangles based on my construction.

If you notice here there's that central loop in the bottom tangle that I can simply twist over and make it into my top tangle. We see that although the notation is different—although the notation is different—the tangles are actually the same. What is surprising, what is revolutionary about Conway's method, is that he came up with a stunning result based on continued fractions. It is a beautiful relationship between shapes and numbers. Let me explain to you first what a continued fraction is before we jump in to see how this number way of keeping track of objects relates to tangles.

Here we see if we're given 4 numbers, 4, −1, 3, and 5, I can create a continued fraction based on these 4 numbers. So far this has nothing to do with tangles. I'm just explaining what a continued fraction of these 4 numbers is. Here's what we do: Given the first number, I take my first number and I immediately make it into a denominator. My 4 becomes 1/4. To this I add my next number. So −1 + 1/4. I take this new result and I immediately make it into a denominator, 1 divided by −1 + 1/4. To this I add my next number 3 and immediately I make this into a denominator, and I add my last number 5.

At the end given 4, −1, 3, 5 as my number sequence, I get the continued fraction as 5 + 1 divided by the quantity 3 + 1 divided by the quantity −1 + 1 divided by 4. This sequence of numbers that we wrote down are written backwards in a nested formation. This is a continued fraction. Of course we can simplify this fraction. We look at the very smallest piece. We start simplifying and reducing using simple algebra and we end up with 4, −1, 3, 5, that sequence gives us an answer of 28/5 as my reduced continued fraction number.

What is Conway's amazing theorem? It states the following thing. If you're given 2 rational tangles, then they are equivalent if and only if their continued fraction values are the same. What does this mean, "if and only if"? This is a powerful statement. What this says is that if you're given 2 rational tangles then you have 2 pictures that are equivalent, then we know that their number values—their values given in terms of Conway's notation—must be giving you the same continued fraction. That's Remarkable! But more importantly, if you're given 2 tangles with the Conway notation, if you can find the continued fraction values and they turn out to be the same, this means that the tangles—the pictures themselves—must be the same.

In fact, what this theorem really is trying to say is: The continued fraction value embodies the shape of the tangle itself. That 28/5 we came up with last time embodies the shape of the 4 −1 3 5 tangle itself. That fraction value holds all that data in it. What a remarkable statement.

Let's look at this example to see what I mean. One tangle is my 5, 1, 4 tangle. Here what I do—again since it's an odd number of objects—is I take my 0 tangle, I twist it 5 times on the right, twist it once in the bottom, twist it 4 on the right. It's great! Now I create this tangle. Then I have my 2, −2, 2, −2, 2, 4 tangle. Here, since it's an even number, I have to start with my infinity tangle, twist bottom, twist right, twist bottom, twist right, twist bottom, and finally twist the 4 to the right. Both of these values, if you work out the algebra, will give you a continued fraction value of 29/6 for the top and the bottom, which means—without even having to draw the pictures—these tangles must be the same.

There must be a set of Reidemeister values that takes one into the other one. Moreover, remember how we can close up these tangles by connecting the northeast and the northwest together, and the southeast and the southwest together, to get knots or links? If the tangles are going to be identical, then their knots and links must also be identical. This fractional value of 29/6 holds all of that in it. In fact we have seen these 2 knots before. The first one was the projection of the knot which had 10 as the smallest number of crossings we can draw. It had crossing number 10.

And the second 1 had the unknotting number of 2. It's the same knot, using 2 different projections, giving us 1 crossing number and 1 unknotting number when we talked about that in a previous lecture.

We've been focusing on rational tangles, those able to be constructed this way in terms of twisting 1 and twisting the other based on this integer sequence. Of course not all tangles are rational. I can take these 4 strands and do anything I want with them as long as they stay within the circle, but there are some great advantages to looking at rational tangles.

The first advantage is that Conway gives us a powerful method of providing notation for such tangles. Can you imagine calling somebody? You can just call them and say, "Oh, do you know what I'm working on? I'm working on the 5 1 4," and they can immediately draw that picture based on the notation you give them. You don't need to explain how this is created in terms of overcrossing and undercrossing. You can communicate information far more efficiently based on this notation.

Second, we have an elegant method of measuring their equivalence. You don't need to worry about Reidemeister moves and the complications that would involve if you have the notation that continued fraction number captures all the data. Moreover it can be shown that all knots and links obtained from rational tangles are alternating knots and alternating links.

Do you remember how we talked about alternating knots before? These are the ones that have an alternating projection where it goes over and under and over and under as we walked around. Some beautiful results already exist for alternating knots and alternating links. The moment we have these rational tangles that we can construct this way then, we know all knots and links that we get from these rational tangles are alternating themselves.

We'll now construct more complicated tangles and try to understand their structure. The design and construction of tangles beautifully lends itself for the arithmetic operations of addition and multiplication. Do you remember we defined how to add knots together, how to put 2 knots together by clipping the corners and touching them up? We never really figured out how to multiply knots. We never thought about what it means to multiply them. For tangles we can do both.

Addition is simply gluing them together, just like knots. Let's take a look. Given 2 tangles, tangle 1 and tangle 2, what I mean by addition is just connecting the northeast to the northwest and the southeast to the southwest of my 2 tangles. And I've created a new tangle, which has these 4 edges or 4 strands coming out of it; that's great!

What does it mean to multiply? Multiply is a little bit more of a complicated operation than addition, but it's an operation we're already familiar with. Multiplication is the flip glue operation we learned earlier. What I'm first doing is that first I'm rotating tangle T_1 and then I reflect, just like what I did before. Once I do this I glue my 2 tangles together. So my 5, 1, 4 tangle is really 5×1 and then I take that answer and I multiply it times 4. So the 5, 1, 4 really encapsulates a multiplication notation for me. Any tangle formed by addition and multiplication is called an algebraic tangle.

Remember if you can just build it out of these integers, we call them rational tangles. But if you take these rational tangles and do more complicated things with them, like addition and multiplication, we call these algebraic because we have extended this notion of rationality into these algebraic operations. Moreover when you close up these rational tangles, and when you close up these algebraic tangles, we get something called algebraic links.

How closely are these related to the algebraic structures we know of real numbers? How are these related to addition and multiplications of numbers that we're familiar with? First of all addition of numbers has the 0 as the identity. Remember if you take any number 7 and if you add it to 0, you get 7 again. Can we say something like this for tangles? It turns out yes.

Let's take a look. If you take 0 + any number A, notice that this equals A, and this equals A + 0. What if I take my 0 tangle, my 0 tangle + my tangle T, I just glue it up, = my tangle T = my tangle T with 0 added to the other side. It's beautiful! My 0 that I knew for normal numbers turns out to be represented now by the 0 tangle here. In multiplication we had something like this also, $7 \times 1 = 7 = 1 \times 7$. 7 has this 1 value, which no matter what I multiply it with gives me that original thing again which is 7.

Multiplication has 1 as its identity as addition has 0. Do we have something like this for my tangle multiplication? Sort of, so let's take a look at why. Here we see my tangle T and I'm multiplying it to the left by my infinity tangle. Remember how multiplication works, I rotate and then I reflect, I glue together and I get my tangle T again. That's beautiful, but what about to the right? Remember $7 \times 1 = 1 \times 7$. It needs to work on both sides.

If I take my tangle T and no matter what I put to the right of it—for example, even if I put the 0 tangle to the right of it—I need to first rotate my tangle T and then I need to reflect it and I glue it and I don't get what I started from at all. Multiplication fails in terms of having this identity. In fact something called associativity—a property that is so obvious we take it for granted—fails under this multiplication.

Let's take a look. Associativity is when you take 3 elements—1, 2, 3, or A, B, C—and you put them together, and associativity says the way you put these 3 elements together shouldn't matter. A × B × C should equal A × B × C. It doesn't matter how we group them together or associate them together. For tangles if I take tangle $T_1 \times T_2$ and then I multiply it by T_3 after I do T_1 and T_2, look what I get with this picture.

First I need to do T_1 multiplication—and I get the following thing with a rotate, reflect, and glue—and then I need to rotate and then glue to T_3. Then the second part I get this other picture. If I do $T_1 \times T_2 \times T_3$. Here I need to rotate, reflect, and glue to T_3, and then I need to rotate, reflect, and glue T_1 to this answer. At the end of the day, the order in which I'm rotating, reflecting, and gluing—in other words the way I'm associating my multiplication itself, something that was obvious for numbers, 2 × 3 × 4—it doesn't matter if you do 2 × 3 × 4 or 2 × 3 × 4, something that we took for granted, turns out to be not obvious at all. In fact it fails for tangle multiplication.

Thus, although we have some algebraic power more than knots—because in knots we only learned how to add things together—now that we have addition and multiplication, we cannot abuse the system. We don't have too much power. Multiplication works but we don't have this identity, this beautiful number 1. Multiplication indeed works, but we don't have something as obvious as associativity itself.

Based on this understanding of tangles, we can finally discuss mutations, what motivated our work from the beginning. There are 3 possible mutations of a tangle that we can do in a knot. They are all mutants of one another. Let me explain this in a picture.

Here you see 1 particular snapshot of a knot—and I'm looking at a tangle, T_1 inside this knot. Here are the 3 different operations that I can do to T_1, and all 3 operations will result in 3 different kinds of knots, and all of them are called mutants of one another. For example, the first thing I can do to T_1 is I can take it, cut this tangle out, reflect T_1 vertically, get this new picture—I want to say reflect it vertically—and I glue it back in. My knot has of course changed because I've ripped T_1 out—a part of it, sort of a little piece of heart of the knot out—I've reflected it vertically and I've glued it back in.

This new object is called a mutant of my previous one. I can also take this T_1, this tangle inside my knot. I can reflect it horizontally and then I can get the answer and glue it back in. This is also called a mutant of T_1. The last thing I can do is I can take this object, I can reflect it vertically and horizontally and if you do this you'll notice that all I've done is I've rotated it 180 degrees. I've taken my tangle, and by doing horizontal reflection and vertical reflection, I've actually rotated my tangle 180 degrees, this is also a mutation of my first one. This is the mathematical way we're trying to capture what happens to these DNA structures.

We're saying if you take this tangle and rip it, and move things around and put it back in, you've mutated the tangle, mutated the knot itself. There are 2 extremely famous knot mutants and they're called the Kinoshita-Terasaka mutants, and here's what they look like. These are the 2 pieces of the puzzle and notice I've taken my first knot—I've taken a tangle from my first knot—and I've reflected it horizontally. When I do this, I glue it back in and now I get these 2 knots which are different, but they're mutants of one another. Unfortunately mutation operations are extremely hard to distinguish.

Remember amphicheirality for knots when we reflected the entire knot? Even here we struggled to find out how 2 knots are or are not the same and the Jones polynomial helped us. Mutation only reflects and rotates not the entire knot, but just a part of it, making it much harder to tell it apart. Even today, they cannot be detected by anything we've done so far. Not even something as powerful as the Jones polynomial can tell these 2 mutant knots apart. Thus this is something we really want to do for DNA analysis from the mathematical point of view. This shows the difficulty of mutation both in mathematics and nature. We see that greater tools are needed to crack this problem.

What have we done today? We've looked at the idea of mutation from a biological perspective, tried to mimic it into math, understand tangles, and the way algebra of addition and multiplication could possibly help us, and basically I've come to the problem: although tangles are beautiful to communicate information, we still don't have enough weapons to solve more detailed problems.

Thank you for joining me and stay tuned.

Braids and the Language of Groups

Lecture 9

[Groups are] one of the most important mathematical structures. They're this algebraic system that keeps track of numerous things. They appear in several areas, not just in mathematics, but in chemistry, in symmetries of molecules, and in theoretical physics and quantum mechanics.

So far, we have been focused on knots and links, but now we move on to the study of braids. Braids are one of the oldest forms of pattern making in the world and are vital to physics and cryptography. They also further our understanding of knots and links.

A braid is a set of *n* strands attached between 2 horizontal bars; *n* can be any positive integer. In a braid, the strands must always flow down from the top horizontal bar to the bottom horizontal bar. Each crossing in a braid is not just keeping track of switching data, but it's also keeping track of over and under information.

Braids are classified based on the number of strands. We can manipulate the braids only between the 2 bars, just as the strands of the tangle had to stay inside the tangle circle. Moreover, just as we joined the corners of the tangle to form links and knots, we can also extend the strands on top of the braid around back of the bar, closing the braid to form links and knots. The trefoil, for example, is a simple 2-stranded braid with 3 twists in it. In fact, every knot and link is a closure of some braid.

Just as we created a language for rational tangles, we can also create a language for braids. Focusing on a braid with 3 strands, we cross the first strand over the second strand; this move is called sigma 1. The opposite move is sigma 1 inverse. Switching the second and third strands is sigma 2, and the opposite move is sigma 2 inverse. We can create braids using these 4 "letters" as a braid alphabet.

The larger algebraic structure that we're talking about is called a **group**. A group is a set of elements and a way of combining elements. Think of the group as a bag, and we need some way of putting 2 elements in the bag together to get another element in the bag. We call this an operation that works on these elements in the group.

A group must satisfy 3 properties: It must have an identity element (*e*); the multiplication operation must be associative; and every element in the bag must have an inverse such that when the element and its inverse are put together, the result is the identity. The integers under addition form a group, but the integers under multiplication do not. Neither knots nor tangles form groups under addition or multiplication, but braids do. We check identity, associativity, and inverses in a braid with 4 strands to show that braids form groups.

Every knot and link that you can possibly come up with is a closure of some braid. ... If you want to know how knots and links work, you can ask the same question to the braids, and if you have full understanding [of] how braids work, it turns out you'll have a really good understanding [of] how knots and links work.

We can measure the equivalence of braids using our language and 3 rules. According to the first rule, sigma 1, sigma 1 inverse = sigma 1 inverse, sigma 1. In other words, if we put 2 inverses next to each other, they cancel out. This is just like a Reidemeister II move. The second rule says: sigma I, sigma I + 1, sigma I = sigma I + 1, sigma I, sigma I + 1. This is like a Reidemeister III move. The third rule does not relate to a Reidemeister I move; instead, it says: sigma I, sigma J = sigma J, sigma I. We can switch any 2 sigmas as long as I – J is greater than or equal to 2. Just by using these 3 rules and working with letters, we can change one word of a braid into another word of a braid and show that the 2 are equivalent. ■

Important Term

group: An algebraic structure given to a collection of elements with a means of combining the elements (composition) satisfying 3 conditions (identity, inverse, associativity).

Suggested Reading

Adams, *The Knot Book*.

Questions to Consider

1. Draw your favorite knot and try to find two different braids whose closures both result in the same knot.

2. Does the set of real numbers under addition form a group? How about under multiplication?

Braids and the Language of Groups
Lecture 9—Transcript

Welcome back and thanks for joining me again. So far we have been focused on knots and links, but now we move on to the study of braids. Braids are one of the oldest forms of pattern making in this world. We all know about and have seen braids, braiding somebody's hair for example. It turns out that braids are also vital to physics and cryptography. As we will see later they also push our understanding of knots and links even more.

What is a mathematical braid? What does it mean for a mathematician to approximate a real-world situation with braids? Let's take a look. A braid is a set of N strands attached between 2 horizontal bars. Imagine you have these 2 bars and you have N strands. N can be any integer you want that's not negative. You can have 7 strands, 6 strands. In this particular case you have 5 strands. The most important thing about a braid is as these strands are attached to the top horizontal bar and they move down and get reattached to the bottom horizontal bar, those strands must always be flowing down. Imagine it like water. The braid strands must always be flowing down. For example in this picture, you see here we have a braid made out of 4 strands, but this is officially not a mathematical braid because this one strand that starts at the very first position goes and loops around, and the water now flows up as it goes down. This is illegal. We always want our strands to be flowing down.

Note that braids are not like tangles in this way. For a tangle we can take those 4 corners—the northeast, northwest, southeast, and southwest—and we can do anything we want in the middle. A braid has this restriction that although it feels like we can do anything we want inside, it must be going from the top horizontal bar to the bottom horizontal bar. Each crossing it turns out is not just keeping track of switching. For example when you have 2 strands coming and one crosses the other one—say the first one crosses the second one—in our minds we can say this is a permutation, they've permuted or switched, that the first and second strands have switched position.

Each crossing is not just keeping track of switching data, but it's also keeping track of over and under information. As the first and second strands switch, you are forced to make a decision. Are they switching over the other one or is the first one switching under the other one? What is exciting about braids to me is that they appear in theoretical physics as equations.

Let me explain what I mean by this by giving you a little historical background. Chen Ning Yang, in 1968, created an equation in the world of physics with quantum field theory, and in this world of quantum field theory the interest is on the forces that are acting between quantum subatomic particles—how these subatomic particles are pulling and pushing and creating a field based on their attraction and repulsion. Chen Ning Yang created an equation that talked about quantum field theoretic interactions. This was in 1968.

Completely unrelated though, in 1971, was Rodney Baxter, who was studying statistical mechanics, another branch of physics. Statistical mechanics deals with a molecular level—not a subatomic level, but a much bigger molecular level of understanding of things like heat and energy and entropy. Rodney Baxter created an equation which dealt with these statistical and mechanical ideas mostly using ideas related to things like modeling ice—the way ice works and the way heat and energy is transferred in a molecular setting for ice. Notice 1 is talking about a big molecular structure—which is huge compared to a subatomic particle structure—and 1 is talking about how the model of ice along with heat and energy works, and the second is talking about forces within the subatomic systems. Remarkably both of these equations turned out to be the same and they're called today the Yang-Baxter equations honoring the 2 physicists who came up with them, and they both can be amazingly depicted by a braid relationship.

Look at this picture. This picture shows the Yang-Baxter equation. This visual picture represents these 2 physical phenomena that are going on. On the left side you have a 3-strand braid and on the right side you have a 3-strand braid. Notice I can take this top strand that's on the left and basically use a Reidemeister move III it feels like and just move it over the crossing to the other side. The fact that I can do this with these strands is depicting a deep physical phenomena going on in nature.

Braids are also the key to topological quantum cryptography. That's a lot of powerful words packed in there so let me break it down for you a little bit. Cryptography encrypts our data on the web and in numerous places. If you are buying something from a company on the internet, you type in your credit card and it encrypts it. One of the most famous encryptions is the RSA system which scientists have now—along with mathematicians—have come up with numerous and far better ways of doing this. Currently, computer technology cannot crack these encrypted codes, thankfully, or else everybody would have access to our special credit card and social security numbers. A topological quantum computer uses particles called anions to form braids, which form logical gates that make up the computer. In other words the braids themselves are keeping track of how logical network systems move inside this computer.

The advantage of a quantum computer—based on these topological braids—is that it can crack almost any encryption method we use today. It is powerful, but so far it's only theoretical. This machine does not exist. It exists in the minds of mathematicians, physicists, and computer scientists. The braids somehow have a way of keeping track of these logic gates and the more understanding we have of braids—not only do we have a capturing of Yang-Baxter—but we have an idea of how quantum cryptography might work one day.

Consider the basic notion of braids. Braids are classified based on the number of strands. If we want to compare 2 braids the first thing we're going to do is to find out how many strands we have to work with. From now on, let's say if we're given a braid, we're only fixing a certain number of strands to work with. We're not going to compare 2 braids of 2 different types of strands. One can manipulate the braids between the 2 bars; in other words given those 2 bars, they can take those strands and move it around as much as I want—but remember what we did with tangles—I cannot take these strands and move it outside the tangle circle. Similarly I cannot take my braid strands and move it above or below my 2 bars. They have to be locked in place between those 2 bars.

This is similar to tangles of course, just like knot leaving that circle. Let's consider some examples of equivalent braids. Here we see 1 braid given by

3 strands on the left side, and by doing a simple Reidemeister II-like move I can make it look like the central 1, or I can take this right strand over here and just push it under all of these other strands and all these 3 braids are equivalent. This is very intuitive as you can see from these pictures. Notice, though, I have stayed within my top horizontal bar and my bottom horizontal bar.

There's an operation that we can do on braids very similar to what we did with tangles. Remember how we took our tangle and took the northeast and northwest part of the tangle and connected them up, and the southeast and southwest and connected them up to form links and knots? We can do the same thing here. We can take my braid which has N strands—in this particular case we see it has 4 strands—and I can take the 4 strands on top and basically extend them all the way around without introducing any new crossings, and come back and wrap in the bottom. They can basically close up one another, and we called this the closure of a braid. Basically we're identifying the top and bottom horizontal bars. We're making them the same thing and we get knots and links based on this procedure.

Let's ask a difficult question. Which links can be formed from braids? We've talked about a lot of knots and links. Which one of those are coming from braids? Are there braids going on behind the scenes that capture the information of the knots and links that we have worked with? The trefoil, it turns out, is a simple 2-stranded braid and it has 3 twists in it, and if you close up this braid and then free itself of the bar restriction, we've just made it into a simple projection of a knot; it turns out this knot becomes a trefoil. Remember the figure 8 was the simplest knot with 4 twists, with 4 crossings. So what if I take this vertical bar and put 4 twists in it and then glue it up and consider the closure? It turns out when I do this you don't get the figure 8 knot. You actually get a link. That's because with an even number of twists the first strand gets connected to the first strand, and the second gets connected to the second, and you have a link of 2 components. In order for me to get the figure 8, I need to do this particular braid which has 3 strands in it the way it's seen here.

If I close this up, I get the figure 8 and I can get the unknot, for example, in this complicated way. It looks pretty intimidating, but at the end of the day you can show just tangles and untangles into the unknot. You can get the Hopf link, for example, like this.

I've shown some examples of how we can get these things and it doesn't seem obvious. For example the figure 8 wasn't that obvious as we constructed this. We had to actually do some work to actually make it into this. Which knots and links form braids? Can every knot you can draw be made from some braid and taking its closure? A beautiful result was given by James Alexander in 1923. Here's the result: Every knot and link that you can possibly come up with is a closure of some braid. In other words, all our questions of knots and links apply to braids. If you are wondering how knots and links work, every one of those guys is coming from some braid information. If you want to know how knots and links work, you can ask the same question to the braids, and if you have full understanding how braids work, it turns out you'll have a really good understanding how knots and links work.

As we create a language for rational tangles we can also create a language for braids. This language is fantastically useful for us to come up with this continued fraction equivalent notation. I'm going to focus on braids with 3 strands right now. Here's the language I'm going to create: First I'm going to take my first strand and cross it over my second strand, and I'm going to call this move the special move that switches the 2 strands to be sigma 1. I'm going to take my second strand and switch it over my first strand. I'm going to do exactly the opposite move I just did and I'm going to call this sigma 1 inverse; I've switched my first and the second strand. The second kind of move I'm going to do is switching my second and third strand; I take my second strand and third strand, take my second and switch it over the third, or I'm going to take my second and switch it under my third or my third over the second.

If I switch second over third I call that sigma 2, and if I switch second under third I'm going to call that sigma 2 inverse. These are my 3- stranded braids with 4 possible operations I've created. You might say, "What if I want to switch my first and third? What if I want to switch them?" Let's try. If you have 3 strands and if you start switching your first with your third, eventually you have to cross your second somewhere to get over there to the third side.

Therefore, you basically cross the second first—in other words I do a sigma 1 first—and then I need to do the sigma 2 move where I switch my second strand and my third now.

This way we see that these 4 patterns—sigma 1, sigma 1 inverse, sigma 2, sigma 2 inverse—keep track of my braids. How do I use this lettering that I've created? We've basically created 4 letters in an alphabet, and the way I create braids based on my letters is to form words in my alphabet.I can use my 26 letters in my English alphabet to form words by putting the letters together: chicken, I use C-H-I-C-K-E-N. What about for braids? I take my 4 letters in my braid alphabet—sigma 1, sigma 1 inverse, sigma 2, sigma 2 inverse—and I put them together. What does it mean visually to put them together? It means stacking one on top of the other one. For example, take a look here. If I have sigma 1, sigma 2 inverse, sigma 1, sigma 1, sigma 2 I'm writing with my new alphabet, using my new letters I've created and writing a word. What word does this then give me in terms of a braid visual picture?

First I put my sigma 1, I stack my sigma 2 inverse under it, I stack my sigma 1 under it, I stack another sigma 1 under it, and I stack a sigma 2 under it. I've created this beautiful braid that just comes from these letters. Since we know Alexander's theorem which says every knot and link imaginable is coming from the closure of some braid, then it seems that we can describe every knot and link as a word, as one of these words made up of these letters. What a remarkable idea.

Somebody calls you on your cell phone; you answer the phone. They ask you to describe the knot and link you've just drawn on the piece of napkin, which is going to revolutionize life as we know it. As you try to explain the over and under crossings, the communication gets lost. Using Alexander's theorem we know that this knot or link that you've drawn can be represented by some braid, and you call them on the phone and you say "I got it." Sigma 3, sigma 2 inverse, sigma 1, sigma 1, sigma 3 inverse. That's perfect. See how beautiful that is to communicate that knot and link, much less talking about the overcrossing and the undercrossing?

There exists a deeper structure to braids than what we've been talking about. What I want to do is I want to pull back and get a bigger picture of this algebraic structure we're talking about. We have noticed a pattern of adding in numbers, in knots, and in tangles, and now we're going to talk about adding in braids—a form of composition, a form of putting it together. This is simply stacking 1 word that you have under another one. Visually you just put 1 letter after the other, you get a word, you take another letter after the other, you get another word. Iif you want to put 2 words together, you just stack them up under each other.

Is there a general framework to do this? Are we talking about addition and putting things together in tangles and knots in 1 way and in braids in another? Is there a bigger picture going on? This bigger picture is an algebraic structure called a group. What a group is, is the most important algebraic structure that I can possibly imagine. It is stunning. It is beautiful, and it shows up everywhere in mathematics. I'll explain what a group is. A group is a set of elements—we're going to call a group G, to be a set of elements and a way of combining elements together. You need 2 things. You need a bag—your group is your bag with a collection of elements in there—and you need some way of putting 2 elements in the bag together to get another element in the bag. We call this an operation that operates on these elements in my group. This group must satisfy 3 properties.

Let's look at those 3 properties. The first property is that my group must have an identity element. In other words in my bag of elements there must be a special element called E—which is the mathematical symbol for identity called E—which acts like my 0 in addition of numbers. In other words I need to find an element in my bag that must exist there, no matter if I take any other element called A in my group. $A \times E$ must equal A. In other words, it's as if E didn't even exist in terms of this operation. $A \times E$ must equal A must equal $E \times A$, $0 + 5$ should equal 5 should equal $5 + 0$. Notice how 0 was my addition operation as an identity. There must always exist an identity in my group. The second thing is my bag of elements, the multiplication operation—the composition operation of elements—must be associative. $A \times B \times C$ must equal $A \times B \times C$. In other words if I take 3 elements in my bag, A, B, and C, and if I say put it together, well it doesn't matter how you associate them. Remember how we got into problems with tangles before in this thing. Here, it must be associative.

Third, every element in my bag—in my bag of elements, in my group, every element I find—must have something called an inverse, its opposite evil twin. Thus when I put these 2 elements together, I get my identity again. For example, for the integers, if I take the number 7 is there another integer that I can add to it to get my 0 element. Yes, there's −7. 7 + −7 cancel out to get my identity 0. Given any collection of element you pick in the group, there must be its opposite element. Given any A, there must be some B, such that A × B should equal the identity E.

Let's consider some sets that we are familiar with and see if they actually have this beautiful group structure. Remember, we need more than just a set, we need more than just a bag of things. We also need an operation that tells us how to put them together. Let's take a look. What about the integers under addition, 7, −7, 3, 4, −17, these are the collection of integers I have under addition. Is this form a group? I have a collection of elements. I know what the operation is, it's an atom. What are the 3 things? Do I have an identity? Yes, 0. 0 is the identity, and is it an integer? It's an integer, it's in my bag of elements. That's great! Do I have associativity? Is 2 + 3 + 4 the same thing as 2 + 3 + 4 or 2 + 3 + 4? Of course it is. That's great! Do I have inverses? We just talked about it. Every integer has an opposite integer that will cancel it out such that you get 0. Integers under addition form this beautiful group structure.

What about integers under multiplication? I have the same bag of elements, but I've changed my operation. That's all I've done. Integers under multiplication, let's see what we have. Do we have an identity, 7 × something is 7. Yes, 1; 7 × 1 is 7. 18 × 1 is 18. That's beautiful too! I have an identity. Do I have associativity? Is 7 × 3 × 4, does it matter how I multiply these 3 elements and it turns out it doesn't, so I have associativity. It's fantastic so far. What about inverses; 7 × what will give me the identity? Remember my identity now is 1. It's 7 × 1/7 which gives me the identity 1. However, 1/7 isn't an integer. It's not in my bag. Therefore, the integers under addition form a group—they have this beautiful structure—but the integers under multiplication do not.

And tangles, what about tangles? We know a collection of tangles we have in our bag, what about adding tangles together? We know that adding tangles together has that identity, the 0 tangle we talked about last time, and it is associative—you can check that quite easily—but it turns out there's no inverse. There's no tangle, so say if you pick 1 out you have an opposite tangle to cancel it out—remember knots when you put them together you get more and more complicated—so tangles don't have inverses under addition of tangles. Tangles under multiplication are even worse. They definitely don't have an inverse, nor are the associative, which we talked about last time.

Who, then, cares about groups? What's the big deal? They're one of the most important mathematical structures. They're this algebraic system that keeps track of numerous things. They appear in several areas, not just in mathematics, but in chemistry, in symmetries of molecules and in theoretical physics and quantum mechanics. We're going to see them in future lectures, I guarantee it. It turns out the braids given N strands form a group. This is absolutely beautiful because nothing else has formed a group. Knots don't form a group. There's no inverse of a knot that cancels a knot out to give you the unknot. Tangles don't form a group. Braids though—this beautiful visual thing we have—form a group. We need to check the basic properties to make sure this is true. I'm just going to show the example of braids with 4 strands. Remember we can only talk about braids with a certain number of strands at a time, and I'm going to stick to braids with 4 strands first. Then you can imagine how this exact set of ideas I'm going to give you will generalize to other collections of strands. Let's check it out.

We need to check 3 things: Identity, associativity, and inverses. In this picture we see that the identity is just a bar with 4 vertical strands coming down. Notice that if I take my identity element that goes straight down with no crossing, and if I add it down here to a braid 1 way, you get the original braid again. The 4 vertical strands simply stretch things. They don't do anything different. Similarly, if I take my braid and put the identity element under it, these vertical lines, they again just stretch things in the bottom. I can make it easily my braid. I have, therefore, an identity element.

What about associativity? Remember I need identity, associativity, and inverse. I have identity, I have an element, my 4-stranded braid that does this, beautiful. What about associativity? In associativity I have braid A,

4-stranded braid—complicated as you want—braid B under it, and braid C under that. I have these 3 elements. If I take my first A and then I take my B braid and I put them together, and then put my C on the bottom, or if I take my A, B, and C, but first I put these 2 first and then I put this on, you see that it doesn't even matter the way I stack. At the end of the day you get this big stack. The order in which I compose my stacks doesn't matter at all. Associativity comes for free, which is fantastic!

I just need to check 1 thing: inverses. But inverses has been my stumbling block. We even had a problem with inverses with integers and multiplication. This is the hardest thing to do. There's nothing called the opposite of a knot. We couldn't take a knot and untangle it by putting another knot next to it. It just got worse, and we couldn't do this for tangles either. We can do it for braids, though. Let's take a look. Here I have my braid. Remember my braids can always be written using my language that I created. If I have my language of sigma 1, sigma 2 inverse, sigma 3, sigma 2 inverse, here's my 4-stranded braid that I create from this language. What is the inverse of this? If I want to cancel things out so I get back to my identity braid—which is my vertical line, that's what an inverse means—I need to cancel it out to get an identity.

What I do is first I read things backwards. Remember when you stack things, the last thing you stack is the first thing you want to cancel. You want to start canceling things inside out from the very center of the braid all the way out. The last thing I stacked was my sigma 2 inverse, but what cancels my sigma 2 inverse is my sigma 2. If I put a sigma 2 next to it, it cancels it out. It just untwists it. A sigma 3 inverse cancels out my sigma 3 that I had before that. A sigma 2 cancels out my sigma 2 inverse before that, and finally sigma 1 inverse cancels out the sigma 1 that I started with. The braid—sigma 1, sigma 2 inverse, sigma 3, sigma 2 inverse—is exactly the opposite, is the inverse of this other braid—sigma 2, sigma 3 inverse, sigma 2, sigma 1 inverse. If I stack my first braid—my original 1—and this new 1 under it, they both unravel perfectly. The inverses cancel out and I get this vertical line. Remember for rational tangles we had continued fractions, which have this notation to measure equivalence of braids. Can we do something for braids to measure equivalence? We have this beautiful way of creating a group structure based on braids, but can we measure equivalence? It turns out we can. Given 2 braids, by words, we can just look at the words to see if the braids are equivalent.

Remember for tangles—these rational tangles—we had these continued fractions and they were enough to encapsulate that information. These words I claim capture everything you want to know about the braid, and we're going to change 1 word into another word that is equivalent using 3 elegant rules. Here's my first rule. My first rule says sigma *i*, sigma *i* inverse = sigma *i* inverse, sigma *i*. In other words if I put 2 inverses next to each other they just cancel out. We already knew this to be true. This is how I got my inverse braid operation in my group anyway. What this means is, if you have given 2 words and you see sigma *i*, sigma *i* inverse next to each other, you can actually erase it out of the equation because it's like a Reidemeister II move. The sigma *i*, sigma *i* inverse says sigma *i* twists this way, and the sigma *i* inverse twists back, and you can just pull it through to a Reidemeister II move. That's my first operation.

My second operation that I can do in terms of these words is the following: Sigma *i*, sigma *i* + 1, sigma *i* = sigma *i* + 1, sigma *i*, sigma *i* + 1. What does this mean? This means if I have 3 words which look like sigma *i*, sigma *i* + 1, sigma *i*, I can actually erase that and replace it with these other 3 words, sigma *i* + 1, sigma *i*, sigma *i* + 1. Visually you have the following braid structure on the left as the first word and the following braid structure on the right as a second word. Notice, these are equivalent because I'm just doing a Reidemeister III move. I can take that strand in the back and the crossing on top and I can just push the strand past the crossing to the other side. When I do it you see that clearly sigma *i*, sigma *i* + 1, sigma *i* on the left picture is exactly sigma *i* + 1, sigma *i*, sigma *i* + 1 on the right. The first move that I had was basically a Reidemeister II move and this power that I have now is a Reidemeister III move. This is basically the Yang-Baxter equation we talked about earlier, in disguise.

There's 1 other move I need. It turns out it's not Reidemeister move I because Reidemeister move I—remember that it had this twist that I had to untwist, but in a braid I can never go up and down—remember water always has to go down, it can never go up—so I'll never see a Reidemeister move I.

I have another move which is not easy to think of. Here's what it is. It says that sigma i, sigma j = sigma j, sigma i. You can switch any 2 sigmas you want to any time as long as $i - j$ is bigger than or equal to 2. The absolute value of those values has to be bigger than or equal to 2. In other words they need to be so far apart that you can switch them. Let's take a look at this picture. Here I have sigma i on the left with some other space in between, some other strands, and sigma j on the right. Remember they're far enough—they're bigger than or equal to 2 away. Thus all this operation says is I can take that crossing of sigma i, I can take the crossing of sigma j, and just literally pull them up like this, just a stretch. If I read the lettering now, I read the top 1 first and then I read the bottom, I and J have switched places.

In fact, let's see what we can do here. This word sigma 1, sigma 2, sigma 1, sigma 2 inverse, sigma 3, sigma 1 inverse, just as a word—let's start using our 3 operations. Well look at the sigma 1, sigma 2, sigma 1. I can just replace sigma 1 2 1 by 2 1 2, that's my second operation. Great, I'm going to do that. My sigma 2 and sigma 2 inverses are now next to each other. I can cancel that out. So let me do that. That's my first move that I can do just on letters. Remember I'm not drawing a picture at all, I'm just looking at the letters. So they go away. Now I'm left with sigma 2, sigma 1, sigma 3, sigma 1 inverse. Let's look though at the last 2 things. Do you see the 3 and the 1 inverse? The difference between 3 and 1 is 2. That's great, that means I can just switch them using my third rule. If I switch them, therefore, I see that 1 and 1 inverse now are in the middle, and they're actually next to each other, and by my rule number 1 they cancel. Cancel that gives us sigma 2, sigma 3.

Just by using those 3 rules and working on letters, I have changed one of my word of a braid into another word of a braid, and I claim that these 2 are equivalent. If you draw the first word, the first long word we started with, you get this picture. If you draw the second last word—the sigma 2, sigma 3 I ended with—I get this picture, and you can simply show that these 2 pictures are identical. Just by manipulating these words I was able to get them to be the same.

In closing, we have studied an extremely important algebraic idea, that of a group. We've entered the world of algebraic topology. We have created this amazingly new language in which to describe knots and links, just being

Platonic Solids and Euler's Masterpiece

Lecture 10

The most famous of all polyhedra are the 5 platonic solids. ... These are: the tetrahedron, ... made up of these 4 triangles; the cube, made up of these 6 squares; we have the octahedron, made up of these 8 triangles; we have the dodecahedron, made up of 12 pentagons; and we have the icosahedrons, made up of 20 triangles.

In this lecture, we move from 1-dimensional objects, such as knots and tangles, to 2-dimensional objects called surfaces. All the objects around us are 3-dimensional, but we observe only their 2-dimensional surfaces. We'll begin by looking at the sphere.

We start with the representations of the sphere with which we are most familiar, polyhedra. Polyhedra are **isotopic** to spheres. The advantage of working with polyhedra is that they have a finite number of flat faces, they have corners (vertices), and their faces meet along edges. The most well-known polyhedra are the 5 platonic solids: the tetrahedron, cube, octahedron, dodecahedron, and icosahedron. Each of these satisfies 2 conditions: all their faces are identical and the same number of faces meets at every vertex. These solids all appear in nature, for example, in crystal formations.

If we try to build these objects by construction, we can prove that we get only 5 of them. For example, if we put 3 triangles together along the edges, we get a tetrahedron; 4 triangles, an octahedron; and 5 triangles, an icosahedron. If we try to put 6 triangles together at a corner, however, the object becomes flat.

The cube has 3 faces meeting at one corner. The opposite, or dual, of the cube is the octahedron. The octahedron has 4 triangular faces meeting at a corner. The 4 for the number of faces and the 3 for the triangle make a 4, 3 combination. For the square, the 3 for the number of faces and the 4 for the sides make a 3, 4 combination. These objects are dual to each other; they look different, but they somehow capture the same data.

An amazing pattern exists for polyhedra, as follows: the number of vertices + faces = the number of edges + 2. For example, the tetrahedron has 4 vertices, 4 faces, and 6 edges: 4 + 4 = 6 + 2. This topological formula is called Euler's formula in honor of its discoverer. It governs the ways these objects can partition, or cut apart, spheres. We can prove this formula operationally by breaking any polyhedron into triangles, removing one triangle, laying the shape out flat, and deleting outside triangles until we have only one left. During the entire process, the value $v - e + f$ never changed. At the end, we get $v - e + f = 2$, which means that it was equal to 2 during the entire process.

An amazing pattern exists for polyhedra. … If we count the number of vertices, edges, and faces of a platonic solid, we always get that the number of vertices + faces = the number of edges + 2.

This formula has applications in chemistry, notably to the molecules called fullerenes. A fullerene is basically a family of molecules formed entirely of carbon atoms, in which each atom has exactly 3 bonds coming to it and the faces must be pentagons or hexagons. In other words, the vertices are carbon atoms, the edges are bonds, and the faces we get after building these atomic bonds are made up of pentagons and hexagons. Remarkably, no matter how we build a fullerene—even using the most complicated structure of carbon atoms possible—as long as it satisfies the 2 conditions of being built out of carbon atoms with 3 bonds at every corner and using only pentagons and hexagons, the resulting shape must have exactly 12 pentagons. The golf ball is an example of this result: As long as a golf ball has divots made up of pentagons and hexagons, not matter how many divots we have, 12 of them exactly must be pentagons and the rest must be hexagons. ■

Important Term

isotopic: A notion of equivalence, the strongest in the world of topology. Two objects are isotopic if they differ by stretching (rubber sheet geometry).

Suggested Reading

Cromwell, *Polyhedra*.

Richeson, *Euler's Gem*.

Questions to Consider

1. Draw three random polyhedra and show that Euler's formula works for all three of them.

2. If we are allowed to have a polyhedron with only squares and hexagons, does Euler's formula provide a relationship between the number of squares and the number of hexagons?

simple letters. Groups are so important. As I said before, we will see them again, not in 1-dimensional setting anymore, but in 3 dimensions in future lectures. Join me then and stay tuned.

Platonic Solids and Euler's Masterpiece
Lecture 10—Transcript

Welcome back and thanks for joining me again. This is our tenth lecture together and for that we move from 1 dimension—worrying about knots and tangles and braids—to now 2 dimensions. We have covered a lot of material in the world of knots and links. Remember how we talked about coloring, to arithmetic of adding and multiplying these objects, from tangles to braids, and the notation of groups, to the powerful polynomials we get from the Jones polynomial, and even to mutations.

Today we move on to 2-dimensional objects, which we call surfaces. Surfaces are the most important of all shapes, I believe, to mankind. This is because our eyes capture and study the surfaces of objects all around us.

All the objects around us are 3-D. This podium is 3-D. I am a 3-dimensional being, but what you see of me is only the 2-dimensional part. You only see the outer shell of what I am. If you'll look around from where you're sitting or where you're watching this you will see everywhere 3-dimensional objects surrounding you, but yet, you're only able to observe the 2-dimensional surface of that object.

The simplest and most prevalent surface in mathematics is the sphere. It is a great test example. It is simple to study, and if we have a good understanding of the sphere, we might begin to get a better understanding of more complicated objects.

This starting lecture focuses on what I believe to be one of the greatest theorems in the study of the sphere. It states as follows: No matter how we cut up the sphere into pieces, there will be a deep relationship between the resulting pieces. Of course I'm being vague here, but the idea is that something is going on about the sphere based on the pieces of the puzzle that you get in forming the sphere.

We just want the outside 2-D shell of the sphere, so every time I mention words like the sphere or a surface, remember we're not talking about the inside regions at all, just the 2-dimensional outer shell—an infinitely thin layer that's coating it. In other words what your eyes are seeing when you're looking at the sphere, that shell of it. We start with the representations of the sphere which we are most familiar with, especially as kids, notably polyhedra.

What are polyhedra? You've seen examples of these before. Here's a cube made up of a lot of flat pieces—kids' toys—that we can play with. Now polyhedra are isotopic to spheres. We talked about this earlier, but what isotopy means again is rubber sheet geometry. If you're given a sphere you could push and pull, no cutting and gluing is allowed. You can push and pull and stretch the sphere and make it into the cube. Similarly you can take a cube, pump air into it and inflate it and make it into a sphere. So these are all isotopic in terms of this rubber sheet geometry.

But the advantage of working with polyhedra is that they have these finite number of faces which are all flat. They also have these corners, the vertices of the polyhedra, and the faces, and these places where the 2 faces meet called the edges. These 3 ways of breaking up the polyhedra into vertices and edges and faces gives us a better way of approaching a sphere, more tractable, because it has to do with numbers again. We can try to count the number of corners, edges, and faces of the polyhedra.

The most famous of all polyhedra are the 5 platonic solids. What are they? These are: the tetrahedron we have made up of these 4 triangles; the cube, made up of these 6 squares; we have the octahedron, made up of these 8 triangles; we have the dodecahedron, made up of 12 pentagons; and we have the icosahedrons, made up of 20 triangles.

What makes these objects special? They are called regular polyhedra, which means they satisfy 2 conditions. First of all, if you notice for any of these platonic solids—any of these regular polyhedra,—we notice that all the faces must be identical. They're all triangles, or they're all squares, or they're all pentagons, and if you notice, all of those objects are regular. In other words this is an equilateral triangle, all the edges and angles are the same. And similarly for the cube, we have all the squares that make up the faces of the cube are the same, 90 degrees, perfectly formed and being identical.

The second thing we notice about the platonic solids is that the same number of faces must meet at every vertex. For example, here we see the octahedron. At this corner, at this vertex, 4 triangles meet, but at every corner, at every vertex, 4 triangles meet. These 2 conditions of being made up of the same pieces of the puzzle that are regular and glued exactly the same way at every corner makes them the platonic solids, makes them regular polyhedra.

As for the name, Plato wrote about them in 360 B.C. in his work *Timaeus* where he associated each of the solids to an element of this world, Earth, air, fire, and water, and the fifth solid possibly being the cosmos itself. Johannes Kepler believed them to embody nature itself. He took Plato's ideas 1 step further. In Kepler's work *Mysterium Cosmographicum*, the Cosmographic Mystery was published in 1600, where here Kepler took these 5 platonic solids and nested one inside the other one, separating each one of them by a sphere.

He took a small sphere. He nested a platonic solid around it. He took another sphere, platonic solid, another sphere, platonic solid, and he kept doing this until he had nested all 5 platonic solids with 6 spheres, the smallest sphere on the inside and then all the other ones between with a sixth on the outside. He believed the ratio of the nesting of these 5 solids bounded by these spheres corresponded in ratio to the then-known 6 planets of the Solar System. His idea was that these 6 spheres that were capturing the platonic solids kept track of the Solar System in a beautiful way.

Kepler later abandoned this idea when he realized that the orbits of the planets are not circles, but they turn out to be ellipsis, which Kepler became famous for. As more planets became available, as scientists were able to discover more of these, this concept of these 5 platonic solids somehow capturing this information went away. For us, we notice that these also appear in nature, not just in keeping track of cosmology and all of these deep ideas in space which it does not turn out to do.

Tetrahedral, cube, and octahedral, for example these 3 platonic solids, appear as crystal formations in nature all the time. In 1904 a German biologist names Ernst Heinkel found radiolarians, which are protozoa having detailed mineral skeletons in the shape of things like the icosahedra, dodecahedra, and others. These shapes, these platonic solids, are somehow intrinsic to nature. They show up very naturally because of their beautiful symmetry.

In fact, outer protein shells of numerous viruses that we know of form regular polyhedra. For example, the deadly HIV virus has the structure of the icosahedron itself. Let us prove why there are only these 5, and we're going to prove this purely by construction.

Let's take a look here. Let's actually try to build these objects by construction and prove that we only get 5 of them. Here's how we do it. We have a collection of different pieces that are all regular. Here're some triangles, here're some squares, here's a pentagon, and a bunch of their friends, some hexagons. Let's see what we can do.

First of all, we know that the sides of the puzzle, the sides of these platonic solids have to be polygons. Here is the smallest polygon there is. You can't find anything smaller than a 3-sided triangle. There's nothing called a 2-sided polygon.

Let's take some of these triangles and see what we can do. Remember the construction. These are all perfectly equilateral regular triangles. These are all regular squares, regular pentagons, and regular hexagons. We have the first condition of regularity met, so the only other condition we need to worry about is the fact that the corners, the vertices, have to have the same number of pieces around each one.

let's take a look. If I start building this we can see that with 2 of these triangles put together—that's not a vertex here yet—I need to add another one. Let's add another triangle here. If I add the third triangle, we get—actually it's sharp enough for it to actually bend all the way around and get this extremely sharp vertex at this point. This formation with 3 at this corner is exactly what gives us the tetrahedron when we close the last 1 up.

That's great! That's how the tetrahedron is formed. We added 3 at this corner. The other possibility we can do is, we can try to add 4 at this corner. So let's try to add 4 and see what we can get away with. If I add 4 at this point, working diligently and faithfully with toys, we have 4 at this corner—

they're all regular, condition 1 is met, 4 at this corner—and you can see it works. It actually locks together in place. If we do this on the other side, we get exactly the octahedron. That's the second object that we can get from this.

Let's try 5 at this corner. So putting a fifth 1 here, locking it into place with 5 of them. Here we get again something that works, and if we complete this entire object, 5 at a corner, it perfectly fills into my icosahedron, the 1 with 20 sides. Let's keep going. Why stop here? Let's keep going. Let's try 6 at a corner. Excellent, perfect, wonderful, my giftedness is in construction it turns out. If I have 6 at a corner I don't have a vertex. I don't have that vertex anymore. It's flat. In fact, it looks like I have these hinged edges, but there's no corner there to speak of. So 6 is not possible, and if I try another one, now this becomes certainly not the corner that we want at all.

It seems the most I can every do with triangles is that I can get either 3, 4, or 5 and those are 3 out of my 5 platonic solids. Great, my triangles are done, and now let's move on to squares, the next size I can worry about.

In squares again, if I take 2 of these pieces of the puzzle and try to fit it together, my squares, 2 isn't enough, I need 3 to actually start building a corner. Let's take my third 1 here. Go this, and we see with the third it works beautifully, and I get actually the corner of a cube. That's another platonic solid. We've done 3 with my triangles. I have my fourth 1 with my cube. Let's now be ambitious. Let's try to get 4 in a corner and if I do this, then you see that 4 in a corner with a cube is again flat. The angles add up to 90, 90, 90, 90 to 360 degrees. Four gives me this flat object. I can't do any more than 4—4 already has this flatness—so I can only get 1 out of this one.

I have 3 from the triangles, 1 from the squares, and now let's try these pentagons. If I try a pentagon, again we need at least 3 to get going and it actually works out that you can actually snap 3 of these together. That's great! If I do this, this is the formation of a dodecahedron, the 1 with 12 sides. Now if I try to put 4 of these pentagons together, which I really want to see how much I can push, you notice that they won't lie flat. In fact, they don't even have a corner anymore; it's an awkward shape. We're done with pentagons. We get 3 from triangles. We get 1 from squares, and we get 1 from a pentagon. Let's keep going. Why not?

Let's try a hexagon, and in a hexagon we can fit 2, we can fit 3—remember we need at least 3 to start getting any kind of corner—and 3 shows that it's completely flat, that the angles add up to exactly 360- degrees. I can't do anything with this, much less 4 isn't going to work out either. There's nothing I get with a hexagon. Any other shape beyond this, also, is completely going to hurt me because of the fact that the angles are too much. Even in hexagons, the angles are too much.

So what have we talked about? We have right now shown in a rough approximate way by construction that there are only these 5 platonic solids that are possible. If we look a little bit closer, however, it turns out that there's more to the platonic solids than we think. Consider for example the cube. The cube has 3 faces meeting at one corner. The opposite—or the dual—of the cube turns out to be the octahedron. The octahedron has 4 faces which are triangles, which meet at a corner. The 4 for the number of faces and 3 for the triangle, it's a 4,3 combination, but this one has 3 faces that are squares. Three faces meeting at a corner, each 1 has 4 sides. This is a 3,4 combination, a 4,3 and a 3,4.

They're dual to one another. They look different, but they're capturing somehow the same data. Similarly we notice that there is duality between these 2 objects. Here we have 5-sided things, 3 meet at a corner, 5,3. Here we have 3-sided things, 5 meet at a corner, a 3,5. Here's this duality we have again for this.

What about the tetrahedron? It's 3-sided, 3 meet at a corner. The 3,3 is the same thing as itself. It's its own dual. Not only can we form these 5 platonic solids and show there are only 5 of them, but there's this duality that exists between them. An amazing pattern exists for polyhedra. It's as follows, if we look at it for the platonic solids in particular. If we count the number of vertices, edges, and faces of a platonic solid we always get that the number of vertices + faces = the number of edges + 2.

Let's take a look. The tetrahedron here has 4 vertices, 4 corners, plus 4 faces, that's 8. The number of edges it has is 6, well 6 + 2 is 8. 8 = 8. That's beautiful! Let's try a cube. The cube has, what do we have? It has 8 corners, 8 vertices plus 6 faces, 8 + 6 is 14. But it also has 12 edges, 12 + 2 is 14. What about the dodecahedron? The dodecahedron here has 20 vertices—we can count them—and 12 faces, that's 32. If you count the number of edges, you get 30 edges, 30 + 2 is 32. There's this beautiful relationship for these platonic solids. Vertices + faces = edges + 2.

This is a beautiful property. Why do these solids have this property? Is it because they're regular or they're highly symmetric? It turns out Leonhard Euler, around 1750, discovers that this relationship—vertices + faces = edges + 2—doesn't just hold for the platonic solids. It holds for every polyhedra possible. Euler is one of the greatest mathematicians of all time, along with Archimedes, Newton, and Gauss.

Euler lived for less than 80 years, but he wrote enough mathematics to fill 74 volumes of work, more than any mathematician ever. He was prolific. Recall the 7 bridges of Konigsberg problem we talked about earlier? That was due to Euler. This formula vertices + faces = edges + 2 is called Euler's Formula in his honor. Notice that it is not a geometric formula. It has nothing to do with the angles, or area, or length of sides, but it's a topological one. It governs the ways these objects can partition or cut apart spheres.

Remember all the platonic solids, in fact, all the polyhedra are approximations of spheres using flat sides. This formula is something about a sphere itself, and thus it's topological. Let's try to prove this formula and show that it always stays true.

Here's the way the proof goes. We're going to begin with any polyhedron possible. Pick a dream polyhedron of your choice, one of these platonic solids or anything you want to come up with. We need to show for this polyhedron you picked—any polyhedron possible—vertices − edges + faces = 2. I'm just going to move the E from one side of the equation to the other, so vertices − edges + faces = 2. Each polygonal face that you can possibly imagine—it doesn't matter whether it's a pentagon or a square or anything you come up with—each polygonal face can be cut up into triangles. I'm actually changing the numbers, though.

Let's see what happens. If a face is not a triangle I'm going to add an edge from one vertex to another one cutting that face up into a triangle. As I start cutting this big face into smaller triangles, into pieces that are triangles, what am I doing? If I add an edge, I've increased my edge value by 1. At the same time, every edge I've added I've increased my face value by 1. Remember what used to be 1 face, by adding this edge, now I get 2 faces. This increases the value of E, but it also increases the value of F for every edge I add.

In the formula V − E + F, remember E goes up and F goes up, nothing has changed. Thus my V − E + F value, if I take my polyhedron, the 1 you give me, and cut the faces up into triangles, I haven't changed this value yet. That's the first thing I'm going to do. What I end up with is a polyhedron made up entirely of triangular faces. I'm going to pick one of those triangular faces, remove it and set it aside, I want you to remember this for later that we've done this. I'm going to take my hand, put it in the triangular face and do a rubber sheet deformation. Stretch this out and lay it flat on the plane.

The faces and edges get deformed, but the count does not change. We have the same number of faces, the same number of edges, and the same number of vertices. Keep in mind this triangular face that I've taken out, for later. Here's what I'm going to do. Since the 1 I have now on the plane—this stretched face of my polyhedron—is made entirely up of triangles, I'm going to start deleting triangles one at a time, and there are 3 possible ways I can do this.

My triangles can come in 3 places. I can have a triangle where one of its edges is on the boundary like follows. If I delete this triangle, what happens? I lose that edge on the boundary—if I've deleted it, I just lose that edge on the boundary, I lost an edge—but I also lost a triangle itself. That's a face. I've lost an edge, I've lost a face, but E and F of opposite sides. That means my value, V − E + F, doesn't change. What happens if my triangle on the outer part of the region—remember how I lost the one with 1 edge on the boundary—what if now I pick a triangle on the outer part of the region with 2 edges on the boundary like this?

If I delete this triangle I've lost 2 edges, but at the same time I've lost that tip, that vertex, and I've lost the triangle itself. My V and F decrease by 2 and E decreases by 2. V and F have the same sign and E has an opposite sign so again perfectly it cancels out. My V − E + F formula does not change if I delete one of these kind of triangles.

There's only 1 other triangle I need to worry about on the outer boundary. Either I have a triangle with 1 edge on the outer boundary I can delete, or a triangle with 2 edges to delete, or the third case if I have a triangle with 3 edges to delete. I delete this edge which looks like this, but here notice 2 vertices are deleted. Three edges are deleted and 1 face is deleted. Thus my V − E + F formula still remains the same. What are we left with at the end? We are left with 1 triangle at the very end which has 3 vertices, 3 edges, and 1 face.

That's great! V − E + F, 3 vertices, 3 edges, that's 1. Remember that triangle that I threw away in the beginning to make it stretch and put down, that's another face. At the end of the day I have 3 vertices, 3 edges, and 2 faces. Thus V − E + F = 2 is what I end up with. Doing the entire process, my value V − E + F never changed. At the end of the day I have V − E + F = 2, that means it must've been equal to 2 during the entire process, and I did this for the polyhedron you gave me, the most generic polyhedron you could think of.

That means it must be true for every polyhedron possible. I can perform this same operational method of breaking into triangles, taking the triangle out, laying it flat, and deleting these outside triangles until I have 1 left. What a stunning result. Beautiful result. No matter how we build your polyhedron, the number of vertices, and the number of edges, and the number of faces have to be in this balanced position where V − E + F always equals 2. This is Euler's remarkable result. We're going to use this formula numerous times throughout our lectures.

Consider the following applications to chemistry based on this formula, notably to the molecules called fullerenes. What is a fullerene? A fullerene is basically a family of molecules formed entirely of carbon atoms. If you take carbon atoms, put them together in certain ways, you get these objects called fullerenes, and these fullerenes mostly appear in the shape of spheres. They were discovered in 1985 by Robert Curl, Harold Kroto, and Richard Smalley, and they cleverly named it after Buckminster Fuller and his geodesic domes that he's extremely famous for.

Why did they name it after these domes? Because these objects as you put the carbon atoms together start taking the shape of these polyhedral structures, these polyhedral domes. Because of this, these 3 scientists were awarded the Nobel Prize in chemistry in 1996 for this discovery. The most famous fullerene is the C_60. It's called the Buckminster fullerene, again to honor him. So what makes a fullerene? They are made up of carbon atoms where each atom has exactly 3 bonds coming to it. If you take a carbon atom as your building block, you have to put 3 bonds or 3 segments coming from this thing.

We know we can build this fullerene based on this structure. Moreover, the faces must be pentagons or hexagons, and this is determined by the chemistry properties. In other words, you have these vertices to be now your carbon atoms, your edges are your bonds, and then the faces that you get after building these atomic bonds—these structures—are going to be made up of pentagons and hexagons.

I want to show you a stunning result. These kinds of fullerenes that you can get based on this property, where every vertex has 3 bonds coming out of it—every carbon has 3 bonds, and you get pentagons and hexagons—no matter how you do this you will always have exactly 12 pentagons in every such fullerene. It's remarkable. It doesn't matter if you start building it right now, if you start building it out of the most complicated structure of carbon atoms possible, as long as it satisfies those 2 conditions of making it out of carbon atoms with 3 bonds coming at every corner and you only have pentagons and hexagons to deal with, every possible shape must have exactly 12 pentagons.

Let's take a look. We begin by letting P be the number of pentagons and H to be the number of hexagons. My goal is to convince you that P must be 12, that it can't be any other number but 12 pentagons. I'm not going to tell you anything about the number of hexagons. That's not what I'm interested in. I want to convince you 12 pentagons is the only thing possible. Let's see what we can do. How many faces do we have in this polyhedral structure made up of carbon? We have the number of faces to be P + H, the number of pentagons plus the number of hexagons. That's the only kind of structure we have here, restricted by chemistry.

What about the number of edges? Each pentagon gives me 5 edges and each hexagon gives me 6 edges. Thus, if I take every pentagon P and multiply it by 5, 5P, and if I take every hexagon H and multiply it by 6, 6H, I get the total number of edges, but I've double-counted because if I take a pentagon and a hexagon they have to glue along this edge. That means from a pentagon's perspective, I've counted this edge and from a hexagon's perspective, I've also counted that edge. I'm double-counting my edges. I see that 5P + 6H = 2E, 2 times the number of edges.

What about the number of vertices? Each pentagon has 5 vertices, each hexagon has 6 vertices. So 5P plus 6H gives me the total number of vertices. Once again, though, I've over counted because if you look at a vertex—remember how 3 things meet because of the bonds, there are 3 of them—it cuts it into 3 faces. That means I'm triple counting a vertex. This vertex is counted from this face. That vertex is counted from this face, and I'm counting this vertex from this face. Therefore, 5P + 6H = 3 times the number of vertices. I'm triple-counting the vertices.

I have 3 formulas now. F = P + H, 5P + 6H = 2E, and 5P + 6H = 3V. If I put all of these together, I know I can use Euler's formula because I've come up with a polyhedron. This is the fullerene I'm working with. Euler's formula says vertices − edges + faces, no matter what kind of polyhedron you have—it doesn't care about hexagons and pentagons, no matter what you have—it's going to be vertices − edges + faces = 2.

What do I know about vertices? Let's start substituting. With 6V, I can plug in my V and multiply this whole equation by 6. Instead of a V − E + F = 2, let me have 6V − 6E + 6F = 12. I know what 6V is. I know what 3V is. That means 6V must be 10P + 12H, let me plug that in. I know what 6E is because 2E is 5P + 6H. Then 6E is 15P + 18H. I'm going to plug that in, and the third 1 I know is 6F. Well an F—the number of faces—is P + H, so 6F is 6P + 6H.

I plug my formulas, my 3 formulas I got from the fullerene, into my Euler formula and I can simplify all of this and I simplify it, notice what happens. The 10P − 15P + 6P cancel out and I'm ending up with just P alone, and my 12H − 18H + 6H all cancel out, and I'm just ending up with no Hs. At the end of the day, I get my new formula P equals 12. What a beautiful result. This says that no matter how we build our fullerene, as long as it's made up of pentagons and hexagons, we'll always end up with 12 pentagons.

In fact the dodecahedron shows an example of this. This is, in some sense, a very special fullerene. It's made up of all pentagons and no hexagons, and we see that the number of pentagons is 12. There's a very special example of this beautiful result. If you look at your everyday golf ball you will notice that these golf balls have exactly the same property we're talking about where there must be exactly 12 pentagons on each golf ball. As long as the golf ball has these divots made up of pentagons and hexagons, no matter how many divots you have—1,000 divots, 100 divots, 383 divots—12 of them must exactly be pentagons and the rest must be hexagons.

We were able to get this stunning result of fullerenes and simple things like golf balls based on a use of Euler's remarkable formula. We have now begun our study of surfaces. We have left the world of 1-dimensional knots and links and moved on to the 2-dimensional world of surfaces. We've started by looking at polyhedra, in specific the platonic solids, and then we stepped back and looked at polyhedra in general and got Euler's stunning formula. We've also proved using Euler's formula, one of these beautiful results with consequences to fullerenes and to nature.

Stay tuned for the next exciting lecture.

Surfaces and a New Notion of Equivalence

Lecture 11

If we had a hard time telling knots apart based on isotopy, surfaces are going to be much harder. They're 2-dimensional objects. Thus, a new concept of equivalence is needed—[homeomorphism].

This lecture looks at 2 questions: What possible shapes could the Earth have been had it not been a sphere? How would we know the shape of the Earth if we were not allowed to leave it? We begin by asking another question: What do we mean when we say "surface"? A surface must look the same at every point. This is called the neighborhood condition. A surface must also be finite in area.

What do we mean by surfaces being equivalent? With knots, we talked about equivalence as being isotopic. We can stretch and pull, but we cannot cut and glue. If we apply the same definition to surfaces, we can see that the power of isotopy is such that it doesn't tell us much about equivalence of surfaces. We will need a new concept of equivalence to tell us whether 2 surfaces are the same or different. This new concept is called homeomorphism, meaning "similarity of form." Two surfaces are **homeomorphic** if we can cut surface 1 into pieces, pull the surface apart based on these cuts, manipulate each piece we choose up to isotopy, then glue the pieces exactly the way we cut them—along the same seams—and get surface 2.

Homeomorphism is a weaker notion of equivalence. In isotopy, we're not allowed to cut and glue. Here, we can cut and glue, but we have to reglue the same way we cut. Under homeomorphism, all knots are the same. We can cut a knot at any point, then stretch and pull to untangle the knot, reglue it, and get the unknot.

What does homeomorphism actually measure? What surfaces are equivalent under homeomorphism, and what is it trying to say? To understand this, we need to consider the difference between extrinsic and intrinsic. Extrinsic is based on what your world looks like if you're an outside viewer. Intrinsic is based on what your world looks like if you live on the surface. Isotopy measures your world the way you would look at it if you left the world. Homeomorphism measures your world the way you would look at it if you lived in the world. With the cutting and gluing, though it looks like you're shattering the world, at the end of the day, since you have to glue the same way you cut, what would once be apart would then be put back together identically. Isotopy is deformation in the extrinsic world, whereas homeomorphism is deformation in the intrinsic world.

Isotopy measures worlds the way you would look at it if you left the world. But homeomorphism measures worlds the way you would look at it if you lived in the world.

Given a surface, we learn that no matter how we partition the surface into pieces, the value we get from the number of vertices minus the number of edges plus the number of faces based on the partition (a value called the Euler characteristic) does not change. In other words, we can partition the surface any way we want into vertices, edges, and faces, and this $v - e + f$ does not change; it's a property given to the surface itself. Thus, we say that the Euler characteristic is a homeomorphic invariant. Using a formula to relate the Euler characteristic to genus (roughly defined as the number of holes in a surface), we can find out which kind of surface we live in. ■

Important Term

homeomorphic: A notion of equivalence, weaker than isotopic. Two objects are homeomorphic if one object can be cut up into pieces, stretched, and reattached along the cuts to form the other object.

Suggested Reading

Adams, *The Knot Book*.

Richeson, *Euler's Gem*.

Questions to Consider

1. Convince yourself that the Euler characteristic of a genus-3 surface is, in fact, –4. Draw some concrete examples.

2. If two shapes are isotopic, is it easy to show that they are also homeomorphic?

Surfaces and a New Notion of Equivalence

Lecture 11—Transcript

Welcome back and thanks for joining me once again. Today we continue our explorations of surfaces. Previously, we talked about spheres as the main focal point of surfaces, the most prevalent and simple surface that mathematicians study. We also considered polyhedra which were isotopic to these spheres. You can use rubber sheet geometry; you can make rubber sheet geometric moves to make a sphere into polyhedra.

We even closed the previous lecture by a beautiful result about fullerenes. We said that any object, any polyhedral shape that you get made out of hexagons and pentagons, if every corner has 3 points meeting together, it will always have exactly 12 pentagons. A normal soccer ball has this property, golf balls have this property, and fullerenes have this property.

In this lecture, we want to move beyond the sphere. We want to motivate it by considering the shape of our own Earth. We all know that the Earth is a sphere, but we ask 2 questions. Question number 1, what possible shapes could the Earth have been had it not been a sphere? Question number 2, how would we know what the shape of the Earth was if we were not allowed to leave it?

Today we can leave the Earth through satellites and space shuttles and view its shape from out of the Earth itself. What if we were stuck on the Earth, like Columbus or Magellan, and must find its shape from within the Earth's perspective? Would we be able to guess the shape of the Earth based on being on the Earth itself if it was not a sphere?

First, what do we mean when we say surface? Last time we considered surface and related it to polyhedral approximations of spheres. But this time we want to step back and get a bigger picture of what surfaces are really about. A surface must look the same at every point. It must, locally around the point that you're standing at, on the surface look like you're standing on the plane. This is called the neighborhood condition.

Here we see examples of objects which fail the surface condition. Let's take a look at 3 specific examples. Consider the disc. It's just a circle with the surface filled in between the circle to give us this disc. This object feels like a surface to us. At every point on this disc, if you look close enough, it looks like it is a plane, a whole 360 degrees of movement. However, if you go to points on the boundary of this disc, it fails this condition. At the boundary of a disc, if you look around, it doesn't look like an entire plane, but only part of the plane.

What about this other example where we see 2 spheres touching at a point, a wedge of 2 spheres? We see that this is not a surface again. At every point, life looks fantastic. At every point, you have a whole plane's worth of information around that point locally until you get to the point where the 2 spheres meet, at that wedge itself. If you look around this point, if you're standing at the point looking around, you notice that the region around you doesn't look like a plane, but it looks more than a plane. It looks like 2 planes meeting at a point. We have more than 360 degrees of information. The disc, the boundary points on the disc, had too little and this has too much.

What about this other example where we take a sphere and attach just an interval, a little line segment touching at a point on the sphere? Again, the points on the sphere look great until you get to the point at this intersection. Here, if you look around, you have 360 degrees of movement on the sphere. You have this extra degree of direction you can travel. In fact, any point on this interval you notice is a 1-dimensional piece of movement. You don't have a whole plane's worth of possibilities.

These are examples of things that aren't surfaces. What is an example of something that is a surface? Look at this picture. Here we have a complicated, stretched out object. It turns out that this is a surface. In fact, it's isotopic to the sphere. I can perform rubber sheet geometry and try to make it as spherelike as I can.

A surface must also satisfy another condition. Not only must it have this neighborhood condition where around you, you have 360 degrees of movement, but the surface—according to the way we want to define it today—must also be finite in area. Although the plane is a 2-dimensional surface because at every point in the plane it, of course, looks like part of the plane—and this happens to be probably the most popular surface used in geometry in high school and college—it fails to satisfy our condition of a surface.

We want our surface to be finite in area and a plane is too big for us to handle right now. It has infinite area. It goes on forever. Here're some examples of surfaces that do satisfy this condition. A sphere is a beautiful surface. It's finite in area and at every point on the sphere, we have a whole 360 degrees worth of freedom. Another example of a surface is something called a torus. A torus is just the boundary of a donut. It's just this ring of an infinitely thin sheet that's surrounding this donut. Remember, there's nothing inside it. It's just the shell of this donut.

This satisfies the condition also. It clearly has finite area because I can hold the torus in my hand. Also, it's a surface because at every point we have a whole 360 degrees worth of freedom no matter where I'm standing on this torus.

Now that we have an understanding of surfaces, what do we mean by surfaces being equivalent? Remember, every time we introduce a new concept we want to understand mathematically what equivalence means. In knots, we talked about equivalence as being isotopic. You can stretch and pull, but you cannot cut and you cannot glue. We can use the same definition for surfaces. We see that 2 surfaces can be equivalent up to isotopy, which means we have this rubber sheet geometry with us. We call these isotopic surfaces.

Let's take a look at some examples of isotopic surfaces. Notice here that there's going to be no cutting or tearing. Here are some examples of things that are isotopic to spheres. I can take an example of the sphere and stretch it using rubber sheet geometry to get this example or this other one. They're all equivalent surfaces under isotopy. What about a surface which has say 3 holes, something that looks like a triple donut? I can take this, I can put my fingers through the center hole and stretch it. Notice I get this surface here in the middle. Or I can take the original surface and pull it in 3 different directions—again, no cutting or tearing—and I get this third object. They're all equivalent under this concept of rubber sheet geometry or isotopy.

Notice that 2 knots are equivalent if they're isotopic to one another. The same kind of definition as what we used for knots and we want to use it for a surface. But, now I want to show you something that's truly remarkable. I want to show you how powerful isotopy is on a surface by looking at this demonstration of a double torus, a surface with 2 holes like a torus, and see what it does.

It's called the clothesline trick. Let's take a look. The power of isotopy we have. What I want to do is take a clothesline and put 1 hole of my 2 punctured torus here, my 2-holed torus, just put my 1 hole here. What I want to convince you of is just by moving this through rubber sheet geometry, by isotopy, I can also take my other hole and put it in through the ring without taking it out of the ring. In other words, I'm going to put 1 hole through this rod and just by moving it around, I'm going to put this other hole here using isotopy.

Let's try this. The first thing I do is I keep the rod fixed. I deform things. This is just Play-Doh so I'm able to deform as much as I can. I can deform things to look in a rubber sheet geometric way, just flipping it around on the other side. Now I'm going to stretch this part, that's the center line, on the left and I'm going to swing over this other part to the right.

Let's see what happens. I'm just stretching my Play-Doh. No amazing results yet. Here I've just stretched it. Great! But, now I'm just going to do more deformations by squishing in this Play-Doh. This is my stretch or compress, but again no tear or cut, so I'm compressing it. As you can see, I'm compressing it even more. I'm compressing it even more. That's great. Notice what I have. I have both of my loops through this rod. What used to have 1 loop through the rod now has both of them through the rod just by this concept of rubber sheet geometry.

This tells me that this geometry seems extremely powerful. How powerful is this? How powerful is isotopy when we talk about it on a surface rather than our knots? Can we take the torus and using isotopy, make it into the sphere? Can we take this torus and just stretching and pulling, can we get rid of that hole in the middle? Our gut says no. Our gut says, somehow that's instinctive. You have to tear the torus itself to get rid of that hole. But, look at this powerful demonstration we had of something that intuitively you would think was not possible, taking this one loop of this double-holed torus and putting both loops in there. It seems we can get away with a lot of stuff in this surface isotopy. It also seems that we're having the same kind of issue we're having with knots.

We're trying to find out which objects are the same. Is the torus the same as the sphere and which objects are different? If we had a hard time telling knots apart based on isotopy, surfaces are going to be much harder. They're 2-dimensional objects. Thus, a new concept of equivalence is needed.

We create a new equivalence called homeomorphism, meaning similarity of form. Remember, our previous equivalence was isotopy, the lens we used to look at surfaces and knots, which is rubber sheet. But, homeomorphism, its definition of equivalence can be defined as follows. Two surfaces, surface 1 and surface 2, are homeomorphic if I can cut surface 1 into pieces, pull the surface apart based on these cuts, manipulate each piece I want up to isotopy, and then glue the pieces exactly the way I cut them, along the same seams. If I can take my first surface, cut it open into pieces, do whatever I want with each piece, and then glue it back exactly the same way I cut along the same seams and get surface 2, then I say that surface 1 and surface 2 are homeomorphic.

This homeomorphism is a weaker notion of equivalence. Remember, in isotopy we're not allowed to cut and glue. Here you are given that ability. But, you aren't given too much freedom. You're not given too much power because although we can cut and glue, we have to reglue exactly the same way we cut it. Your freedom, although more than isotopy, is limited. How weak is homeomorphism in telling things apart?

Under homeomorphism, all knots are the same. Why is that? Because I can take a knot, cut it at any point I want, and then I'm just going to do isotopy and untangle the knot and I'm going to reglue it again and get that unknot. I'm going to reglue the same place I cut it, that's all my rule is, as long as I glue back the way I cut. Thus I can take any knot and using this concept of homeomorphism, I can make it into the unknot. All knots, according to homeomorphism, are the same. Thus, this is not very exciting in 1 dimension. That's why we stuck to the concept of isotopy.

As we will see, this is extremely useful in 2-D. Since we have a more complicated object to study, the 2-dimensional surface, we will need a weaker notion of equivalence to tell things apart. What does homeomorphism actually measure? We know what isotopy measures. It's measuring up to rubber sheet pulling. We have an intuition for that. But, what does this cutting and regluing actually do for me? What surfaces are equivalent and what is it trying to say?

To understand this, we need to consider the difference between extrinsic and intrinsic. Let me explain. Extrinsic is based on what your world looks like if you're an outside viewer, if you could leave your world. To us, my surface is an extrinsic way of thinking about it and isotopy is sensitive to this. If I have a surface, if I pull it and stretch it, from an extrinsic viewpoint, I see how that's changing. It's not changing that much.

But, let's consider it from an intrinsic perspective. Intrinsic is based on what your world would like if you lived on the surface. If I take an object and perform a homeomorphic change—if I cut, rearrange, and then reglue the same way I cut—how would my world look from an outside perspective? My world would look drastically different. A knot, which is quite complicated, would extrinsically all of a sudden become the unknot. My world has completely changed from an extrinsic perspective. But, if I do a homeomorphism move, if I cut and make it into the unknot, how has my world changed intrinsically?

If I lived on the knot, if my whole world was the knot itself, then from an intrinsic perspective, what would once be knotted from an intrinsic perspective, I would just be able to walk around the entire circle whether it was knotted or whether it was unknotted. It would be the same thing. Isotopy measures worlds the way you would look at it if you left the world. But, homeomorphism measures worlds the way you would look at it if you lived in the world. This is because the cutting and gluing, though it looks like you're shattering the world and manipulating it, at the end of the day, since you have to glue the same way you cut it, what would once be apart would then be put back together identically.

This is the difference between homeomorphism and isotopy. Thus, isotopy is a deformation in the extrinsic world whereas homeomorphism is deformation in the intrinsic world. Again we ask, what possible shapes can the Earth have been? Recall that we ask a surface to satisfy the finiteness and the neighborhood condition. A sphere is a possibility; a torus is another possible shape that the Earth could have been. We call this a surface of genus 1 since it has 1 hole. This concept of hole of a torus is called a genus in math.

We can take 2 tori, 2 different torus'—we call toris plural—cut out a hole, say a small disc, in each one of them and glue the boundaries of them together to get a surface of genus 2. When we get the surface of genus 2, this is exactly like the addition of knots. Notice how we snipped a little piece of each of my strand of knots and connected it up. In a similar way, we can take little pieces of my surfaces, 2 tori, torus 1 and another torus 2, and cut these pieces, glue it together, and get a new one.

In this process, we can cut another ring of my genus 2 surface, attach another torus to it, and get a genus 3 surface. We can continue this process of getting higher and higher genera, higher and higher genus surfaces. Again, we're left with 2 questions. Are we actually getting new surfaces as we glue these tori together? Or are they somehow the same under homeomorphism? It looks like we're building things with higher and higher genus—we are—but, is there a way under homeomorphism that I can cut, rearrange, and magically get the sphere? Maybe everything just magically becomes the torus. How do I know I'm actually getting something different?

If they are different, how do we know it's genus? How do we know it's genus if we look at it intrinsically? In other words, if you have a genus 1 surface and a genus 2 surface, can you tell by living in the worlds themselves, either in this world or in this world, which surface you're living in? Both of these questions get answered by Euler, the same person who gave us the Euler's formula last time.

The power of Euler appears again in the world of surfaces, not just in the world of polyhedron. Our motivation is inflating a polyhedron from the last lecture and making it look like the sphere. We see that V, the number of vertices, E, the number of edges, and F, the number of faces, will actually be some of the topology of the sphere, not the geometry of the polyhedra itself. Thus, for a given surface, S, along with a partition of cutting the surface into regions, now the surface S can have as many holes or genera as you want. You can take a surface of genus 4 or a surface of genus 7. It doesn't matter.

We're going to cut the surface into regions. Each region is made up of vertices, edges, and this one nice connected face. Given this, we define this mathematical number called the Euler characteristic of a partition of a surface. If somebody gives you a surface and somebody tells you how to cut the surface, a partition of the surface, we create something called the Euler characteristic. This is the number of vertices minus the number of edges plus the number of faces of this partition. We're not talking about spheres anymore. It's of any surface possible.

It's a remarkable result which generalizes our ideas from the polyhedron case that the Euler characteristic of any surface does not depend on the partition you pick. In other words, if you are given a genus 4 surface, no matter how you cut it up, the number of vertices minus the number of edges plus the number of faces will always remain the same. The Euler characteristic is something that belongs to the surface itself and not to the partition of the surface.

Consider 2 different triangulations of a surface. Let's call it triangulation T_1 and triangulation T_2. I want to present to you a rough proof of why this has to be true, why Euler's characteristic, this Euler characteristic we came up with, does not depend on the triangulations. It's built into the surface itself. The first thing I want to do is I want to overlay these 2 triangulations on top of each other and build a new object called T_3.

We can build this new triangulation, which contains both my original triangulation, T_1, and my second triangulation, T_2, overlaying it. We add extra edges to make sure that this new object is made up of triangles itself. What I want to convince you of is that I'm going to build this new triangulation T_3 from T_1, my original triangulation, using a 4 step process. At each step, I want to show you that the Euler characteristic isn't really changing. Eventually, I want to start at T_1 and build my object called T_3.

The first thing I do is take my triangulation T_1 and I look at all the extra vertices from T_2. Remember, triangulation T_3 is made up of T_1 and T_2 put together. I take out all my vertices from T_2 that happen to intersect on edges of T_1. I throw those in there. Every vertex I throw, which intersects an edge, if I put that vertex, my vertex number increases. But, my edge gets split into 2 pieces so my edge number increases. It is vertices minus edges. Since they're opposite sides, it cancels out beautifully.

The second thing I do is add in the extra vertices that I have which didn't happen to land perfectly along these edges. When I do this, not only do I want to add an extra vertex that lands on one of these faces, but I want to throw in an edge that connects it from a previous vertex. Remember, a vertex can't just land in the middle of nowhere without an edge touching it. Thus, I throw in that vertex and I throw in an edge, some edge that you want to connect to that vertex.

What have I done? I've increased my vertex count by 1 for each new vertex I throw in, but I've increased my edge count by 1. Vertices go up, edges go up. Perfect! It's right on the mark that V- E + F hasn't changed yet.

The third thing I do is add in the rest of the edges of my T_2 triangulation. I throw in those edges. Notice all the new edges I throw in don't come with any new vertices because we already have all my vertices that happen to land on edges and all my vertices that happen to land in the middle of faces. I don't have any new vertices to add in, so I'm only going to be adding extra edges connecting previously existing vertices. Now if I do this, if I add every extra edge, it cuts a face that I have into 2 pieces. Thus, my edges increase each 1 I add. But, my face count increases each 1 I add. Edges go up, faces go up, $V - E + F$ stays perfect.

The last thing I need to do is throw in any extra edges I have to form a triangulation of T_3. When I do this, for every face that isn't a triangle already, I can throw in extra edges to make it into a triangulation. But, by throwing extra edges in, no new vertices, I increase my edge count. But, each edge I throw in cuts the face into 2 pieces so I increase my face count. Thus, I've started at T_1, my original triangulation. I've built this new triangulation called T_3, which is made up of T_1 and T_2 superimposed, along with any extra edges needed to make it into a perfect triangulation.

I started at T_1, made it to T_3, and I showed that the Euler characteristic here, the vertices, edges, and faces, is the same as the Euler characteristic here in terms of vertices minus edges plus faces. But, I could have started it not just at T_1. I could have started it at T_2 and built T_3 the exact same way. Thus, the Euler characteristic of the T_1 partition and the Euler characteristic of the T_2 partition are both equivalent to the Euler characteristic of the T_3. This means that they must be equivalent to each other.

Thus, the Euler characteristic of T_1 and T_2 have the same value. The vertices, edges, and faces, when you look at $V - E + F$, do not depend on the partition. It depends on the surface itself. What does this mean for us? It means that the Euler characteristic is fundamentally related to the surface. It is a homeomorphic invariant. Two surfaces are homeomorphic if we can cut 1 up and reglue the same way. Notice that when you cut things up and reglue, you haven't changed the Euler characteristic. Why is that? It is because when you cut it up and reglued, you put things back exactly where you found them. Only when you look at it from the outside, from the extrinsic perspective, do things look twisted or changed. Intrinsically, it hasn't changed.

Let's look at some consequences for this thing. Consider the Euler characteristic of the sphere. The Euler characteristic of the sphere, as you know here, only depends on its partition. Let's cut up the sphere in any way we want. I'm going to cut it up like this, where I have 3 triangles on top and 3 on the bottom. Here I have 5 vertices, 9 edges, and 6 faces. Therefore, $V - E + F$ is 2. Perfect! It doesn't matter how I cut it up, I just proved it depends on the object itself. No matter how you cut up the sphere in terms of vertices, edges, and faces, you will always get 2.

In fact, this is Euler's formula when we talked about partitioning polyhedron. $V - E + F$ is 2 is exactly this. But, my friends, we can do so much more. Consider the Euler characteristic of a torus. Here, I'm going to partition it this way. If I partition it along this way, cutting up the torus like this, I get 12 vertices, 24 edges, and 12 faces. The Euler characteristic becomes 0. If you choose to cut up the torus you have in your own way, you will also see it doesn't depend on partition. It depends on the surface itself.

What about for higher genus surfaces? We can keep trying this for other things by designing something that partitions our surface, but there's a beautiful trick we can do. Let's take a look. We noted that a genus 2 surface comes from 2 tori glued together. The Euler characteristic of the genus 2 surface must be based on the 2 tori and the gluing. The Euler characteristic of my first torus is 0, of my second torus is 0, but when I cut out those 2 triangles and I glue it, what happens? I've taken 2 triangles out, the insides of the triangles out, so I've lost 2 faces. But, I've also identified what I used to have—3 vertices and 3 vertices, 6 total vertices. By gluing, I only have 3 left over.

I've lost 3 vertices in the process. Similarly, I used to have 6 edges. In my gluing, I've lost 3 edges. By doing this gluing, this addition of tori, I've lost 2 faces, 3 vertices, and 3 edges. My net loss for my Euler characteristic is −2. This means that the Euler characteristic of my genus 2 surface is 0 from the first torus, 0 from the second torus, plus a − 2. My Euler characteristic total for my genus 2 surface is −2. We can continue this pattern, adding more and more of these tori handles to my surface. As I increase the value of genus, I keep adding these tori. Every time I do this trick, I keep losing −2 from my gluing.

I have a formula for relating Euler characteristic in genus. The Euler characteristic of a genus G surface is $2 - 2G$. When the genus is 0, when there is nothing, I get the exact example of the Euler characteristic of being 2, my sphere. When the genus is 1, I get the Euler characteristic is 0. When my genus is 2, I get Euler characteristic is −2. It's a beautiful formula that works because of the way we built these surfaces up. Thus, we can find out which surface we live in, one of the questions I asked before, by simply calculating its Euler characteristic.

If you live in a surface and without leaving the surface—finding a space shuttle or a satellite to understand—if you're in the surface, you simply need to partition the surface which you can do living in it. Then you could just count the vertices, edges, and faces, plug it into this formula, and you can find the genus of the world that you live in. How beautiful!

Let me close by looking at a genus of a very complicated structure. The genus of this structure is something that I cannot figure out just by looking at it. Does it have a couple of holes that are going through and is it 3, is it 1? Here's what we do. The first thing we do is count, since it's already partitioned into pieces, the number of vertices, edges, and faces. In this particular piece, I see that I have 24 vertices, 42 edges, and 16 faces. If I add these values together according to my Euler characteristic formula, $V - E + F$, I get that the Euler characteristic is -4.

But, I know from my previous formula that the Euler characteristic and genus are related. It's $2 - 2G$, which means the genus of the surface must have been genus 2. We have answered both of our questions from the beginning. We can tell surfaces apart by their Euler characteristic and we can find out the genus, a global result of our surface, from just local data. Even without leaving the world you live in, we can find out its genus. What a remarkable thing we can do—just intrinsically we can understand this concept using this idea of homeomorphism, a new concept of equivalence of surfaces.

Next time, we continue our adventures of surfaces. We're looking at more complicated examples now that we have built a foundation to stand on. Stay tuned.

Reaching Boundaries and Losing Orientations

Lecture 12

If you can tell me the surface's orientability—whether it's orientable or not—if you can tell me how many boundary components it has, and if you can tell me what its Euler characteristic is, then you have completely understood everything you need to know about surfaces.

In the last lecture, we saw how the Euler characteristic, which is a local phenomenon, helps us understand a global property—the genus of a surface. We also saw that if 2 surfaces are homeomorphic, they must have the same Euler characteristic. The Euler characteristic of a sphere is 2, of a torus is 0, and of a genus-2 surface, –2. In this lecture, we'll see how to build any surface of any genus by simply gluing together polygons. We'll also learn that there is more to surfaces that just their genus. In particular, we construct surfaces that are non-orientable. These objects have the interesting property that they have only one side. Finally, based on these ideas, we will be able to classify every possible surface imaginable.

In 1884, **Felix Klein**, one of the fathers of topology, developed a method of building surfaces from polygons. We see, for example, how we can construct a torus from a square. In fact, there is a pattern to these constructions, according to which we can build a genus-*g* surface by gluing a 4*g* polygon.

Having used polygons, which have boundaries, to build genus-*g* surfaces, it's natural to consider generalizing the notion of a surface to include things with boundary. Removing interior faces of any surface results in surfaces with boundary components. A classic example is a disc, which we get by removing one face of a sphere. In general, we can give a genus-2 surface 3 boundary components by taking 3 pieces off the surface. Considering this surface, how does the Euler characteristic change with respect to boundary? Every time we remove a face to get a boundary, we lose one value in the Euler characteristic. The Euler characteristic of a genus-2 surface is –2. Thus, the Euler characteristic of a genus-2 surface with 3 boundary components is –2 – 3, which is -5. We can use this information to show that 2 surfaces that are twisted in space differently are actually the same.

Surfaces have another property besides genus and boundary components—orientability. We say a surface is orientable if it has 2 different sides and a surface is non-orientable if it has only one side. A surface is orientable if it has 2 sides that can be painted with 2 different colors, such as a torus or a disc. The Möbius strip is the most famous example in mathematics of a non-orientable surface. Just as we constructed all orientable surfaces by gluing polygons, we can construct non-orientable surfaces by gluing polygons. From the square, we can build the Möbius strip, the Klein bottle (which can only exist in 4 dimensions), and the projective plane.

According to a theorem developed by a number of mathematicians, if we know these 3 properties of a surface—its orientability, its number of boundary components, and its Euler characteristic—we can classify every surface that exists in any dimension possible. ■

Name to Know

Klein, Felix (1849–1925): Klein spearheaded some of the pioneering relationships between algebra and topology. He also showed us how to obtain all surfaces from gluing polygons.

Suggested Reading

Adams, *The Knot Book*.

Richeson, *Euler's Gem*.

Questions to Consider

1. What do you think is the least number of triangles needed to be glued to form a torus?

2. Try to construct a genus-3 surface by gluing the edges of a 12-sided polygon in the right way.

Reaching Boundaries and Losing Orientations

Lecture 12—Transcript

Welcome back and thanks for joining me again. We have, thus far, been studying surfaces. Today, we're going to explore the wildest of surfaces. In the previous lecture, we showed possible models of the Earth, of different genus surfaces. We studied the classic sphere and then we went to genus 1, the torus, genus 2, and higher. We showed how this was related to something called the Euler characteristic, this amazing collection of numbers that were given to each particular surface, 1 number for each kind of surface.

This was such an important concept that I want to review it just for a little bit before we move on because we will need it again today. We showed how Euler characteristic, which is a local phenomenon—something that you can get by counting the vertices, edges, and faces where you're standing as you walk around the surface—helps understand the global property, the genus of the surface itself. The small local count gives you global data.

Given a surface, we learned that no matter how we partition the surface into pieces, the value we get from the number of vertices minus the number of edges plus the number of faces based on your partition, this value, which is also called the Euler characteristic, does not change. In other words, you can partition your surface any way you want into vertices, edges, and faces and this $V - E + F$ does not change if it's a property given to the surface itself. Thus, we say that the Euler characteristic is a homeomorphic invariant. It does not change based on homeomorphic type.

If 2 surfaces are homeomorphic, that means they must have the same Euler characteristic. Why is that? Remember what homeomorphism was. It's taking a surface, cutting it into pieces, ripping the pieces apart, doing whatever we want to the pieces, and then gluing exactly the way we found it again.

Since you cut it, rip it apart, do whatever we want, and glue it the same way, all the vertices, edges, and faces that we originally drew on the surface to cut have been reglued exactly the way we found them. The number of vertices, edges, and faces do not change. Thus, if 2 surfaces are homeomorphic, they must have the same Euler characteristic. In other words, if 2 surfaces have different Euler characteristics, they cannot be homeomorphic to each other.

We computed last time that the Euler characteristic of a sphere was 2. The Euler characteristic of a torus was 0. The Euler characteristic of a genus 2 surface was −2 and thus, since these are all different, we know that these are all different surfaces. Since their Euler characteristics are different, the surfaces are different.

This lecture does several fantastic things. First, we show how to build any surface you want of any genus from gluing together simple polygons, just flat polygons that we understand really well. Then, we show that there is more to surfaces than just their genus. Genus is 1 thing to talk about, but there is more that we haven't even touched upon. In particular, we construct surfaces which are non-orientable. These objects have the interesting property that they have only 1 side to them. We'll go into more detail later on.

Finally, based on these ideas, we will be able to classify every possible surface imaginable. Given any surface—whether you can draw it or it shows up in nature or shows up in the mathematical world—we will have complete understanding of it based on our ideas we learn today.

We have talked about surfaces of genus, G, and shown its relationship to the Euler characteristic. We have this formula that the Euler characteristic of a genus, G, surface is 2 − 2G. In 1884, Felix Klein, one of the fathers of topology—topology was this branch that was starting to grow from geometry in the 1800s and the start of the 1900s—this was the time that Felix Klein made his move. He was a brilliant topologist and he developed a beautiful method of building surfaces from polygons.

In order for us to understand his method, let's begin by considering the square and what life would be like if we lived in the square, if the square was the intrinsic world in which we lived. Note that this is a surface that is finite, that's one of the conditions we wanted. It is also a surface with boundary, the 4 edges of the square, these are the places where you can go and actually fall off the square. Remember this fails the neighborhood condition that we wanted a surface to have.

What happens if we start to manipulate the surface? Consider the classic example of a cylinder made from a square. Here we see a square and I'm going to do a mathematical labeling trick, which is drawing arrows on the 2 sides of the square, the left and the right sides. I'm going to have them point the same way and I'm going to call and label these arrows A. What that means is that the left side and the right side are identified; they're actually the same arrow. If I take the sheet of paper, the square—the left and the right side are the same—I can actually roll up that sheet of paper and make the sides touch. Those 2 edges become 1 and by doing so, I get a cylinder.

We have formed the cylinder from this square. We use this notation of using arrows on the edges of the square throughout these lectures. This new object I've created by gluing the left and the right sides of the square together is a new world with different properties than that of the square. Remember, in the square there's this whole square boundary of possibilities where you can fall off, where you have a place where the world doesn't have an entire 360 degrees of freedom around it. Once you glue it together, now you have a cylinder.

With a cylinder, you can walk entirely around the cylinder beautifully and now they're 2 separate components of boundary. There's the top circle and the bottom circle, which weren't connected at all. Previously, in the square, the entire boundary was connected.

If you think about it a little bit, about this property of the cylinder coming from the square, you notice that this is exactly what's happening in the game *Pac-Man*. Have you ever noticed in *Pac-Man* that *Pac-Man* goes from the left side of the screen to the right and when he comes to the right side of the square that you play the game on, he doesn't die or fall off? He just reappears on the other side. He goes, comes to the right, and reappears on the other side because the *Pac-Man* is being played on a cylinder.

We can also consider another identification of the square, not just the left and the right sides. We can identify the top and the bottom sides as well. Here we see the square. I'm going to identify the left and right sides with an up arrow, call it A, and I'm going to identify the top and bottom sides with a right arrow and call it B.

Let's first start identifying the right and the left sides. I glue them together. I get the cylinder exactly as before. Now I have more to identify because the top ring of the cylinder and the bottom ring of the cylinder are the same thing. What I can do is actually twist the cylinder, just stretch it open—because it's deformation, this rubber sheet geometry of topology I'm in—and glue these 2 rings. If you notice, as I take my cylinder and start to twist and start to glue, right before I glue I need to make sure that the arrows are pointing the same way when I glue it. If you look carefully, you will see that, as you're about to glue, the arrows are pointing the same way and I can glue this thing. What I end up with is a torus.

Now we have constructed a cylinder from the square. We have also constructed a torus from the square. The property of this torus coming from the square is encapsulated in the game called *Asteroids*. In *Asteroids*, you're the spaceship. When you go to the right side of the screen, you come back on the left exactly like going this way on the torus. If you go on the top of the screen you come back in the bottom, exactly like going this way on the torus. *Asteroids*, the game that's played on a square computer screen, is actually played in real life, mathematically, on a torus.

What about other genus surfaces? We've built this torus; can we build more complicated ones? Here we see this object, which is an octagon, a perfect octagon, and I have labeled the octagon as follows. I have arrows pointing in certain directions of 4 kinds of labeling: A, B, C, and D. If I glue these objects along these arrows exactly the way I've defined them in the direction that they're pointing—just like I glued it for the torus, the directions match up perfectly—I will end up with this object, a genus 2 surface.

It's not immediate, like the construction of the torus, getting it from the square. But, you can see here, if you can actually cut this out and try to make it with a little bit of flexibility in which you're working with, you will actually get this genus 2 surface. In the genus 2 surface, notice that I've labeled the 4 rings. Each ring corresponds to my A glued together will form 1 loop, my Bs, my Cs, and my Ds will form all 4 different loops meeting at this one point. It's a beautiful thing and this was Klein's contribution.

It seems like there's a pattern here. We have constructed the octagon, this genus 2 surface with an 8-sided thing and we construct a genus 1 surface with a 4-sided thing. Can we do this in general? There is a pattern. This does work; in general, you can build a genus G surface obtained by gluing a 4G polygon, a polygon with 4 G sides. Why does this work?

Let's take a look at this octagon in a bit more detail. If I take my octagon, slice it at a certain position, and cut it into 2 pieces, I'm going to consider that slice and cut and I'm going to label that edge E. Of course, remember, cutting and gluing are illegal because I'm changing the structure. But, I'm cutting and I'm going to glue exactly the same way I cut. Homeomorphically, they are identical. Let me cut along this line, pull it apart, and label those things as E. At the end of the day, I have to glue the Es back together and get exactly what I started from. I'm not changing the underlying topology.

When I cut it along E, I'll have 2 pieces; each 1 is a pentagon. I can move the E around and make it into a ring. You end up with a square with a little extra ring, that's my E. You have 2 squares with the labeling that you get originally from your octagon now transposed on each of the squares. If you look at each of the squares separately, you see that each square is basically this genus 1 surface, this torus. My first square gives me a torus, my second square gives me a torus, and this little E cut that I made is really 2 boundaries of torus. It's tori with little holes in them.

What I do is glue the 2 boundaries together because I need to glue the Es back together again. Remember, I cut them apart the first time. When I do this, I get a genus 2 surface. Each 4-sided piece that I'm adding on gives me an extra increase in genus. In this way, we can get any surface we want.

Take a look at this genus 3 surface made from this 12-sided polygon. Here, we have this 12-sided polygon, all my labeling, A, B, C, D, E, and F. After I glue all of these objects together, I get this beautiful genus 3 surface along with these boundary components with As glued exactly in the right spot making the A boundary disappear. The Bs gluing and identifying together making the B boundary disappear, on and on and on, and at the end of the day I get a genus 3 surface with no boundary.

There is no place you can fall off because all of the boundary of my 12-gon has been beautifully identified. We have been exploring surfaces of genus G made from polygons.

Let's push the frontier a little bit more. A natural situation to consider is to generalize the notion of what a surface itself is, to include things with boundary. Remember the 2 conditions we started off with last time. We said a surface has to be finite and a surface cannot have boundary. Every point needs to look the same. But, now the boundary seems to be a useful thing to talk about because we have used these things with boundaries to build these beautiful genus G surfaces.

Let's bring in boundary in to our picture. Here we have 2 kinds of points on our surface. Consider a disc. A point in a neighborhood around the disc can have a 360 degrees of freedom or a neighborhood of a point on the boundary of the disc can look like part of the plane, not an entire 360 degrees. You have 2 collections of points. The points that have the entire 360 degrees of freedom and those that don't. The latter set of points, the ones not having 360 degrees, are called the boundary points of the surface, places where you can literally fall off and die.

Why do I make such a big deal of this? It's because the shape of the Earth was once thought to be a surface with boundary. Sailors were warned not to go to close to the boundary of the surface because you might fall off and die. The boundary is a place you have to be careful about because of the topology, how the world feels, completely changes. Instead of having 360 degrees of freedom, you don't anymore and you might leave that surface.

Taking any surface and removing interior faces of the surface results in surfaces with boundary components. Let's take a look at some examples. A classic example is a disc which we get by taking a sphere and removing 1 face. If I remove a face from a sphere, take the interior of the face out, then I'm left with a boundary, a place where I can fall off. This is because you could have continued that on to the rest of the sphere, but now I've removed that part. I can actually put my fingers into that boundary and make it open because of rubber sheet geometry and lay it flat and I get a disc.

A disc is basically a sphere with 1 face removed. If I want to take a sphere with 2 faces removed, then that's simply a cylinder. This is because if I remove this face and if I remove this face, I can put my hand through both of these holes that I've created, these boundary components, and just stretch it out a little bit and I get the cylinder. We can get the cylinder by taking spheres and removing boundary components. Or, in general, we can take a genus 2 surface as you see here and give it 3 boundary components by taking 3 different pieces off.

Let's just consider this genus 2 surface with 3 boundary components. This is one of the reasons mathematicians are hesitant to use the word holes unless the context is very clear. Because here if I say, consider the holes on this object, what do I mean? Do I mean the holes of the genus? Because that is a hole in the surface or do I mean a hole of the boundary component because that's also a hole on the surface. That's why we say genus and boundary component to be really clear. And if the context itself makes it clear, we sometimes do use the word holes. But, we do need to be careful.

How does Euler characteristic, a property we spent a lot of time talking about last time, change with respect to boundary? Consider a genus 2 surface with 3 boundary components, exactly this picture we have. Each boundary component, what has changed for the Euler characteristic perspective? Each boundary component has lost a face. Remember, I've removed the interior of a face out. I've kept all the vertices and edges; the boundary is still there, but the inside is gone.

Thus, every time I remove a face to get a boundary, I have lost 1 value in my Euler characteristic. In order to get a surface without boundary, I need to put that face back on again, which mathematicians call capping off. I'm capping that ring off again to remove that boundary. The Euler characteristic of a genus 2 surface is −2. We know this. Thus, the Euler characteristic of a genus 2 surface with 3 boundary components is −2 minus 3 because I've lost 3 faces, which is −5. We can compute the Euler characteristic of these objects just by looking at the number of boundary components and the genus itself.

Let's look at another example. Consider these 2 surfaces. I'm going to call it surface X and surface Y. What are the genera of these surfaces? What is the genus of surface X and what is the genus of surface Y? If you look at surface X, it's easy for us to tell the genus. It's designed that way, from our extrinsic world. We're looking at the surface from the outside, so we say ah, it has a genus 1 and it just has that 1 boundary component there. Let's just cap it off just to make sure we have no boundary components. Then, if you cap it off, it has no boundary components. We see that this is a genus 1 surface after it gets capped off with no boundary.

We fully understand what X is about. We know its Euler characteristic since it's a genus 1 surface with no boundary, it's Euler characteristic is 0. But, remember it has the original X, it has that boundary. Therefore, we need to lose 1 face so the original Euler characteristic for X, it's Euler characteristic turns out to be -1.

What happens to surface Y? For Y, we actually need to do some work. In the extrinsic world, we're extremely confused as to what this surface looks like. You see that it's a surface because you can walk from the bottom region and you have this twist and then you go to the top region and then you can twist back to the bottom region. Instead of understanding from an extrinsic way what's going on, let's compute the Euler characteristic first—because that's the thing we can actually do, locally we can do that process to get global data.

Let's cut the surface into little strips and regions so we can get a handle on stuff. By cutting it into strips and regions as follows—if I make those cuts—I see that the number of vertices is 12, the number of edges is 18, and the number of faces is 5. Thus, I know that the Euler characteristic of this surface is -1. That's great! But, -1 is the Euler characteristic of the surface. What is the original surface, what is the genus that the surface really encapsulates?

Let's walk on the boundary. Take a point on the boundary and just start tracing it up. As you trace all the way, you see that you come back to where you started from. You've covered every bit of the boundary of the surface. In other words, this surface has only 1 boundary component stretched like this. It's a surface of Euler characteristic -1 with 1 boundary component. If I add that boundary component back on again, through homeomorphism, I can cap it back on. Then, my Euler characteristic increases by 1 because I've capped an extra 1 on so my Euler characteristic is 0.

It's an Euler characteristic 0 surface and it has no boundary now because I've capped it. That has to be the torus. We see that the right surface, surface Y, is a torus with 1 hole missing. The left surface, surface X, is a torus with 1 hole missing. They're the same thing.

If you look at this picture, you realize—how is that even possible? These surfaces are the same thing, but they're twisted in space differently. X, the way the picture is drawn, focuses on its genus, that's what it's honoring. The boundary on X is kind of thrown to the side—the boundary is important—whereas in surface Y, the surface focuses on its boundary. That's what dominates the beauty of surface Y and the genus is somewhere hidden. You have to work hard to find it.

As extrinsic creatures, we're far more comfortable with surface X because we see the genus immediately. But, if you lived on the surface, either X or Y, your life wouldn't be different at all. They're the same thing.

So far, we have seen surfaces having genus and surfaces having boundary components. There is one more property surfaces can have and it's called orientability. We say a surface is orientable if it has 2 different sides and a surface is non-orientable if it only has 1 side. What does this mean? A surface is orientable if it has 2 sides that can be painted with 2 different colors.

Let's look at the torus. I'm going to paint the outside of the torus red and imagine the torus is made up of a very, very thin layer. If I paint the outside of the torus red, I can actually paint the inside of the torus blue. There's this 2 coloring to this torus, the outside of the torus and the inside. I can't see the inside because the outside completely dominates it.

Let's take a look at a disc. I can paint the top of the disc red and I can paint the bottom of the disc blue. You can see that the 2 colors never meet except along the boundary. Since the torus has no boundary, there's no place you can fall off the torus, the colors never meet. On the disc, the 2 colors meet exactly along the boundary. If I lift the disc up, you see where the red and the blue meet perfectly.

Consider this other surface, what we just looked at earlier, where we computed the genus to be a 1, a surface of genus 1 with 1 boundary component. Here you can see that the bottom region I can color red and the top region I can color blue. As I'm walking from the bottom region to the top region, I have this twist. As I twist, I go from the top blue here and it twists under. The blue and the red are perfectly in mesh. As I walk back, it twists back. It's red on the bottom and it twists under to be red and the top has to be blue. Thus far, all our surfaces, all 3 of them here—and every surface we've seen so far—has been orientable.

We say a surface is non-orientable if it has only 1 side. How is this possible? Let's take a look. Consider the Möbius strip. It's the most famous example in mathematics of a non-orientable surface. It's a strip of paper with a twist that you put to glue back together again. It's the square we saw earlier except that instead of gluing it to get a cylinder, I just put a little twist and then I glue it. If you see this picture again, notice, let's start painting my surface red. As I paint, as I continue to paint, as I walk around, it becomes red. I twist and I keep painting red on the inside. I twist and I paint red all the way in the back and I twist back and I keep painting red. In fact, if I keep doing this the entire surface becomes red. There's no place for blue at all. The entire surface is red. There's no place for blue. There are no 2 sides, there's only 1.

Consider another example. Look at this surface here. It looks like just what we talked about earlier with a little extra twist on the right side. Let's start painting it red on the bottom region. If I paint it red on the bottom region, then my twist, since I'm looking under the surface I have to paint, it becomes blue, that's great. If I keep painting it, then it twists back onto my red region on top, great.

If I go all the way to the left, and if I twist, then it has to be painted blue now. But, this is the region that I had started to paint red in the first place. How can I paint this blue if it was red? That means I must only have 1 color. I can't paint a region that's blue that's already red. Thus, I get in trouble, thus the surface is non-orientable. We see that these are 2 examples of non-orientable surfaces. It's impossible for us to use 2 colors on these without having a contradiction so only 1 color is possible.

Non-orientability is very counterintuitive for 2 reasons. First, it is only a global phenomena. Notice that if you stand at a surface at any point, even if it's orientable or non-orientable, life looks the same. It looks like you can walk around and that you can always paint your world red around you. But, if you walk around the entire surface, that's only when you get this contradiction. The Möbius strip looks fantastic at every snapshot. It's when you come back all the way do you get this contradiction that you're in this non-orientable world. It's a global phenomena for us to grasp. Unlike the Euler characteristic, which happens to be a local phenomenon.

Secondly, this non-orientability has drastic consequences. Let's imagine you're a left-handed person and you're walking around the surface. You're left-handed all the time and your friends are right-handed. As you leave your friend on the Möbius strip and come back to them again, making this entire loop, it turns out that you are still left-handed. Nothing about you has changed, but all your friends appear to be left-handed to you. They were right-handed, but now they look left-handed. To them, you were left-handed and now you appear to be right-handed because your hands have lost orientation when you come all the way back around.

Just as we constructed all orientable surfaces from gluing polygons, we can construct non-orientable surfaces from gluing polygons as well. Klein pointed this out again. Just from the square alone, we can get the following 3 non-orientable surfaces. The first one is the Möbius band that I talked about earlier, which is the evil twin of the cylinder. It's the exact same square, the same A labeling of my 2 sides, except I'm going to switch one of my side arrows. When I do this, my gluing, I try to glue A to A, but the arrows don't match if I have a straight gluing. I have to twist a little bit to glue again.

In the second part, I have my square again to build another surface. Now my arrows don't match for my As on the right and the left, but the top ones have my Bs and they do match. First, let's start by gluing my Bs together. When I glue my Bs, I get an elongated cylinder since my Bs have matched up. As I try to glue my As together, they don't match to get the torus. I need to take my object and put the 2 inside it, make it intersect itself and go inside it to try to match up. This is called a Klein bottle. It is a non-orientable surface and it is somehow the evil twin to the torus.

This Klein bottle cannot exist in 3 dimensions because the surface has to intersect; you need 4 dimensions to understand this. We're going to talk about 4 dimensions in future lectures. Thus, let's keep this in mind.

The last object we can get is called a projective plane. It's an entirely new beast that many of us haven't seen before at all. Here, I switch my orientations for A so they're matching up in opposite directions and my arrows, my orientations for B, are also going in opposite directions. I have a Möbius strip, when I glue my As together, and then I need to glue my Möbius strip to itself. By doing this, I get this extremely complicated, extremely irregular object in 3 dimensions. It exists beautifully in higher dimensions if you want to think about this, but this is a projective plane.

It seems like we keep introducing new concepts—genus, boundary, orientability—will this end? It seems like surfaces are getting more and more complicated. We started with The Euler characteristic and genus showed the relationship between them. Now, we're talking about boundary and orientability. There is a beautiful theorem that talks about classifications of every surface possible. It is one of the most amazing theorems in mathematics. It was built not by one person, but by several mathematicians working over decades.

It was based by the works of Klein, of Möbius—who himself is a mathematician in topology—of Dane—who we'll see later—of Haggard—who we're going to see in 3-dimensional objects—and Raddow [ph], ranging from the 1870s to the 1920s. It is based on the equivalence of homeomorphism where we say that every surface that you can draw, every surface that exists in 3 dimensions, or any dimension possible, if you're talking about a surface, do I have a solution for you.

It can be classified by 3 properties: 1, its orientability; 2, its number of components; and 3, its Euler characteristic. If you can tell me the surface's orientability—whether it's orientable or not—if you can tell me how many boundary components it has, and if you can tell me what its Euler characteristic is, then you have completely understood everything you need to know about surfaces.

Remember the Euler characteristic and the genus are somehow the same. They're built into one another. The genus and the Euler characteristic are somehow exchangeable. The boundary components, orientability, and the Euler characteristic tell us everything. Thus, for any given surface, we know exactly what it is. Such a result does not exist for any higher dimensional objects at all. For 3-dimensional objects, 4-dimensional objects, mathematicians do not know how these objects work in full classification. We fully and completely understand surfaces up to homeomorphism. It is a beautiful, beautiful result.

With this lecture, we have a full understanding of all types of surfaces possible. The orientability, boundary components, and Euler characteristic tell us exactly where we stand. Moreover, we have learned from Klein that all our surfaces, orientable or not, can be built simply by gluing edges of polygons.

In the next lecture, we're going to apply this knowledge of this powerful machinery of classification of surfaces to talk about knots. How can surfaces help us understand knots better? Stay tuned.

Knots and Surfaces
Lecture 13

What is an algorithm? This is just a machine; you feed this machine your knot projection, and it spits out for you from the machine the surface whose boundary is the knot diagram you gave it. This is Seifert's amazing contribution.

In this lecture, using the classification of surfaces and invariant ideas for knots, we will try to associate surfaces to knots. Given a knot, our goal is to find an orientable surface whose boundary is the knot. For example, we see that if we shade the regions between the strands of a trefoil, we get a 2-dimensional surface with boundary, but orientability seems to depend on the particular projection. Are we relegated to a case-by-case study to answer this question, based on trial and error and considering different projections?

In 1934, mathematician Herbert Seifert came up with an algorithm to create an orientable surface whose boundary is the given knot. Given any knot projection, we begin by orienting the knot. Then, we look at the crossings and, regardless of whether it's an overcrossing or an undercrossing (positive or negative slope), we split it down the middle and pull the crossing apart. Replacing every crossing with a split leaves us with a collection of disjoint circles. We shade each of these and use them as the surfaces to build from. The last step, to go from these 1-dimensional objects to the 2-dimensional object we're looking for, is to attach strips of surfaces to these circles to form the complete surface. If the original crossing was positive, we put a strip that has a positive crossing on it; if the original was negative, we put a strip that has a negative crossing on it. The result is an orientable surface whose boundary is the original knot. We see this algorithm work for projections of both the trefoil and the figure 8 knots.

Based on this procedure, how do surfaces help us better understand knots? Each of our orientable surfaces has one boundary component. If we input a knot projection into the algorithm, the output is an orientable surface with one boundary component, the knot. If we cap off the boundary component, we get an orientable surface without boundary. From the previous lecture, we know that genus or Euler characteristic completely determines this surface. If we can associate to a knot this genus of the surface, it will be a knot invariant. We will have something associated to a knot that's the genus of the surface we get from building the knot. Unfortunately, it turns out that this is not the case. We can, however, say that the *least* genus of any orientable surface bounding the knot is an invariant. Unfortunately again, there are knots for which the minimal genus surface cannot be obtained by the Seifert algorithm for any projection possible.

Here is what the official definition of the genus of a knot is: The genus of a knot is the least genus of any orientable surface bounding the knot.

Earlier, we were unable to prove that the crossing number of knot 1 + the crossing number of knot 2 must equal the crossing number of knot 1 + 2. We can, however, prove the theorem that states that the genus of knot 1 + the genus of knot 2 must equal the genus of knot 1 + knot 2. This theorem answers a question we asked in one of our earliest lectures: Can 2 complicated knots be added together to form the unknot? Can we have subtractions of knots? By this theorem, we are guaranteed that this cannot happen, because the genus of the 2 knots must be added together to get the new genus. If knot 1 and knot 2, somehow, when we added them together, gave us the unknot (genus-0), then the genus of knot 1 and the genus of knot 2 must both be 0 to satisfy the formula. Thus, knots 1 and 2 would have to be unknots themselves. ■

Suggested Reading

Adams, *The Knot Book.*

Richeson, *Euler's Gem.*

Questions to Consider

1. For your favorite knot, use Seifert's algorithm to find an orientable surface whose boundary is the knot.

2. Show that if a knot has genus 1, then it must be a prime knot.

Knots and Surfaces

Lecture 13—Transcript

Welcome back and thanks for joining me again. So far we have completely classified surfaces. Now remember, surfaces appear all around us. Even though we live in a 3-dimensional world, what we see, what we perceive are just surfaces of objects. Thus if we have a good understanding of surfaces, the usefulness of this shows up in nature all the time, the usefulness of this is important for scientists, engineers, and mathematicians.

We have used colors, numbers, and polynomials to try to distinguish 1-dimensional objects, knots, and links many lectures ago and today we're going to bring our understanding of surfaces and its classification to help us attack the problem about knots from a higher-dimensional perspective. We have, from the previous lecture, a complete classification of surfaces, and for our invariance, we have for knots and links associated values of yes and no, for colorability, numbers—like crossing number, unknotting number on the writhe—and polynomials, like the Jones polynomials to knots.

Today using our classification of surfaces and our invariant ideas for knots, we're going to bring them together to try to associate surfaces to knots. Instead of giving it a number or a polynomial, we're going to give the knot an entire surface and see what we can do. Given a knot, our new goal is to find an orientable surface whose boundary is the knot.

Let's think about that. Given a knot, our new goal is to find an orientable surface whose boundary of that surface is exactly the knot that you've given. Remember, we want to worry about orientable surfaces because when we got into the non-orientable case—like the Klein bottle or the projective plane—things got not intuitive at all. It left our sense of comfort zone, so let's stick with the orientable case, and this is a classic problem that mathematicians have struggled with.

Let's consider the trefoil. How can we build a surface whose boundary is this trefoil? One thing we can do is we can just start shading the regions between the strands of this trefoil, and as we do so for this particular projection, we get this surface. Notice it's a surface, it's 2-dimensional with boundary. What is the boundary of the surface? Where can you fall off on the surface? It's everywhere where these lines are, and as you walk around these strands, you see the boundaries exactly the trefoil, what we started off with.

We have done what we wanted to do, create a surface whose boundary is the knot, but we wanted orientability. Let's just check to make sure this surface is orientable and then we'll be set. This beautiful coloring, the shading trick, of these knot projections seems to give surfaces for us. Let's check orientability in this case.

Let's start by coloring this particular region, this particular region to my surface, red and let's walk around. If this is red, and I want to be orientable, then as it twists I need to reveal the underside of the surface, which must be blue, that's great! I'm going to twist again and make this red, that's great. Now I twist back and make this blue, but this was red to start off with. Thus, I can't shade it 2 different colors—there's only 1 color I can use—which means the surface is not orientable. Although we have a surface whose boundary is the knot, it's not an orientable surface, it's not a nice surface that we can really work with and understand.

Let's try another trick. Let's take the same projection that we have and look at the complement of the shading. Instead of shading the regions we started off with, let's shade the other regions that we missed. Here's what it would look like. It looks like a Halloween pumpkin. Look what I've done, I've shaded everything but the regions I shaded earlier, but I wanted a finite surface with finite area. Notice that it's this sphere with this trefoil carved out in the middle of it. Let's now check if this surface is orientable.

Is it orientable? We can color the entire sphere red, and as I come to the central region of this pumpkin like object, I need to twist, which means it becomes blue in the middle. Then as I leave and go back to the sphere again, it twists back and becomes red. Anytime I'm on the outside it's red, I come back in it's blue, it works beautifully. I can actually use 2 colors. Let's see if the boundary of this object is the sphere, although the entire object is a sphere except for this place, which is exactly the trefoil. We see it is an orientable surface with boundary to be the trefoil, just like we wanted.

What happens if I pick a different projection of the trefoil? Remember how we started with a simple elegant 3-branch projection? Let's pick a different one. Let's take a look at this. We've seen this surface before.

Let's color the left side of this trefoil projection red and the right side blue. As I go from the left to the right, I go from red and I twist over to blue, and I go from blue and I can twist over to red again. It is orientable. This surface we've seen from the previous lecture is a surface that's homeomorphic to a genus 1 surface, a torus, with 1 boundary component, which is exactly the trefoil. We've found this orientable surface on this case with this projection and another orientable surface over here with this projection. That's fantastic!

Let's try it for another knot. Let's see what we get. For this figure 8 knot, which is our next complicated kind of knot because it has crossing number 4, we see that if I start shading at 1 region, let's start with the left side. I'm going to shade red, and as I walk around I need to twist to make it blue, and I need to twist again to make it red on top, and then I twist again. As I twist this last 1 I need to make it blue, again we started with red and now it's forced to be blue, it's not orientable. Great, it's not orientable, it wasn't over yet last time, let's try this complement trick, this pumpkin trick.

I'm going to take my figure 8 knot, put it in the same way I did before except I'm going to draw a sphere around it. I'm going to take this sphere, carve this figure 8 knot out, and here I'm going to shade the entire sphere red, and as I walk in from the bottom it becomes blue, and then I twist again to become red, and I twist to go back into the sphere, and it becomes blue, but it was red in the first place, this is not orientable either.

It turns out for the figure 8 knot, for this particular projection my quick fix of finding the surface of just shading it 1 color and then maybe, if that didn't work, looking at the complement of this pumpkin trick, if that didn't work, then maybe is there a different projection of the figure 8 I need to use? Are we just relegated to a case-by-case study for this based on trial and error and considering different projections? It turns out that this is not the end of the problem.

The power of a Seifert algorithm is introduced. Let me explain to you what I mean. Herbert Seifert, the mathematician, in 1934 came up with an algorithm to create an orientable surface whose boundary is the knot you give it. What is an algorithm? This is just a machine, you feed this machine your knot projection and it spits out for you from the machine the surface whose boundary is the knot diagram you gave it. This is Seifert's amazing contribution.

Let's start by understanding what Seifert algorithm does. The first thing we need to do is given any knot diagram, given any knot projection, you begin by orienting the knot. Let's take a look. Here let's start with the trefoil because it's my simplest case to work with and we already know an answer to the trefoil. Let's see what this algorithm does for us.

Here's my trefoil projection as given and I've picked an orientation. It doesn't matter what orientation I pick, I've just picked one. The first step was to pick the orientation, the second step is to look at every crossing. Remember we always point our arrows up at every crossing like we did before for when we considered the writhe. Here we point our arrows up. What I'm going to do is I'm going to just split, regardless if it's an overcrossing or an undercrossing, regardless if it's a positive slope or a negative slope, I'm still going to split it right down the middle and I'm going to pull these crossings apart. I want to remember this crossing information for the future. Don't throw it away, just keep that on the side of what really happened, but I'm going to cut every crossing apart.

What happens if I do this at every crossing? If every crossing I've replaced it with a split, it leaves us with a collection of disjoint circles. It would look like, in this particular case, 2 circles, all the crossings in the middle were split right through the center. What do we do with these circles? These circles, by the way, are called Seifert circles because they're coming from Seifert's algorithm. I take these circles and I shade each one of them and I get discs. These are the surfaces I'm going to build from.

This is what Seifert did. He takes your projection, he cuts it up into circles, he shades each one of them, he gets a surface, but now this is not the surface you're looking for. They're just a collection of discs. The last step to go from this 1-dimensional object to the 2-dimensional object that we want is we need to attach strips of paper to these circles. We need to attach strips of surfaces to these circles to form the completed surface.

Here are the ways we can do it. If you have a crossing like this, then the strips could look, for example, like this where there's a twist in your 2-dimensional strip, or if your crossings look like this then here's another twist to this 2-dimensional strip going this way. What happens if we put these strips back on? The way we put it back on depends on the crossing information. Remember how we took our crossings whether it was positive slope or negative slope and split it? Now we need to go back and remember which it was in the first place.

If it was a positive crossing, we would put a strip that had a positive crossing on it, and if it was a negative crossing, we'd put a strip that had a negative crossing on it. In this particular trefoil picture we would get a surface with 2 discs on either side, and in the center we would have to remember the crossing information and attach these strips. When I do so, I get a strip in the bottom, a strip in the middle, and a strip on top attached perfectly, and the crossing information that I originally had in the diagram you gave me is now preserved again.

I have a surface, fantastic. Its boundary is the knot, perfect. Is it orientable? We've seen the surface before. This is the surface we've just looked at earlier. We can color the left side red and the right side blue, and as we twist, we go from one to the other perfectly.

Let's consider some examples of this procedure. We've talked about how this works for the trefoil, but trefoil we already had an understanding of, our old shading trick actually worked. The classic figure 8 knot, which did not have an orientable surface, is a more exciting object to think about. Let's take a look at what that would look like.

Here we have my figure 8 knot with an orientation, the first thing I do for Seifert's algorithm, for Seifert's machine. The first thing I do is draw these arrows, and then I look at every arrow, and make sure they're both pointing up and I slice it—if they're both pointing up I slice it—right through the middle and have this vertical split every time. In this particular case I need to slice it here, here, here, and here. When I do this, I end up with 1, 2, 3 circles.

These circles, if you notice, there's on in the bottom and there's 2, 1 inside the other. These are actually nested circles. One is nested inside the other 1, but what that visually means is 1 disc remember we have to shade these Seifert circles and make them discs to go from 1 dimension to 2 dimension. These discs are actually sitting on top of one another, so there's this big disc on the bottom and another one is floating right on top. It's a 3-dimensional way of thinking about what's going on on this flat 2-dimensional paper.

Let's actually take a look now to see how we would glue these surfaces back. When we glue the strips back to get our full surface, it would look something like this. Remember our crossing information. At each one of the 4 places I cut I have to reglue my strips again. Instead of showing it to you in terms of a picture, let's actually build it and see what happens. These are just pieces of discs and pieces of strips that I can glue the way I had exactly to get the crossing information I started with.

Let's look at this demo. Here's what I have. I have the central Seifert circle, which I've shaded it in, into a blue region and there is another circle right here that's actually nested, but remember in 3 dimensions here, the surface is actually floating on top of this. All of my circles have 2 sides. Just to make sure whether this is going to be orientable or not remember I need to check that everything is 2-sided and my coloring works out perfectly. That's 2 of my circles.

My third circle is over here, and I'm going to shade this one red. Here's what happens. I have 1 circle, 2 circles, 3 circles, now we need to attach these strips exactly the same way I did. I've shown you what that picture looks like visually in terms of what's on the screen, but let's try to build it. Here I take a strip and I glue it to this part of my surface and I want to make sure the colors match up to make it orientable, and instead of gluing I'm just going to use a bunch of staples to hold it together.

Notice the twists. I need to twist it this way and notice, as I twist it this way, this is where the crossing information appears from the figure 8 knot. So I twist it this way, now I need to glue it after I make this one complete twist, I need to glue it to this object here, that's exactly what my figure 8 knot tells me to do. Here's my twisted version, but notice so far the coloring works out beautifully. It's blue here and as it twists it becomes blue to the one under it. That's perfect! That's one of them.

The second thing I need to do is I need to make an object that glues from this part to this part with a strip. Let me start here. I start with this strip here, and now I need to twist it this way, exactly the way the figure 8 wants me to do it. And as I twist it this way I need to glue it under here. This is the tricky stapling part, and I glue it under here, great. Notice here's the 3-dimensional part of this. It's actually nested above this object.

I have this third object over here, that's in the front. That's the third circle I have to work with and I have 2 strips again. One strip attaches from this circle, which I glue here, and it attaches to this one, and here the twist looks like this. Here I have this object and this thing twists over this other one. That's great. I have this object right here with another strip I'm going to attach, and this strip twists this way. Notice how the crossing comes in, this crossing comes over this to get to the figure 8 knot that I'm building. Push that in a little bit, and here's my surface.

Here's my entire surface. It's a flat 2-dimensional surface that's in 3 dimensions, and notice the coloring works perfectly and if you trace the boundary of the surface, as you walk around the boundary of the surface, this thing comes here, it twists down and it reappears here. Here's my boundary and I'm tracing my boundary out. You get exactly the figure 8 knot that we wanted. Notice it's orientable. I've made it so. Here's blue and the blue, although there's a twisted strip, stays blue perfectly and the blue goes to the blue. Here the blue again twists under here. Remember it was red to begin with here and it comes back and becomes blue again.

This is my surface. You can actually build it at home to say that this is Seifert's algorithm. It's a physical constructive procedure and not a theoretical procedure at all. The classic trefoil projection also has a different Seifert procedure working on it.

Here we have this trefoil projection, remember how previously we used the Seifert algorithm for a different trefoil projection. Here's the classic 1 with the 3 weaves that gave us the name trefoil. If I take these 3 crossings we get from the trefoil after I orient them, I cut the crossings, these vertical splits for each one of them, I get 2 different surfaces, remember this nesting again is one on top of the other one, and I get these 3 strands. I can glue these 3 strips up exactly the way I wanted to and get again a surface.

Is the Seifert surface I get here orientable? We see it for the figure 8 knot. We actually built it for the figure 8 knot. But what about for this particular projection of the trefoil? Well we can see that the top region can be colored blue, and then as you twist, it becomes blue on the inside, but it becomes red all the way on the outside. In fact, this particular projection we get gives us the surface that the Seifert algorithm produces, which is exactly that Halloween sphere-like object with the trefoil cutout. That's exactly this if you do some rubber sheet geometry deformation.

We see that the algorithm produces surfaces whose boundary is the knot, and in all the cases we've tried, we see that it always becomes orientable. We have something that works. But why must they always be orientable? It's easy to convince ourselves that they're always going to produce the boundary of the surface being the knot, because we cut the surface of up into pieces and then we reglue exactly the way we cut. Of course the boundary is going to be the knot you started off with, but why must they be orientable? What is the reasoning?

Let's take a look. Here we see an example of why it must work for the figure 8 knot. We built this so we know it works, but here's the reason why going on behind the scenes. Recall that our knot begins with an orientation as you see here. Notice every time I cut this into pieces, the Seifert circle inherit the orientation from the knot. Each Seifert circle gets its own orientation based on the orientation of the knot itself. It depends on the way we cut right down the middle.

If the circle is clockwise oriented, let's color it blue here and here. If it's counterclockwise oriented, let's color it red, like here. By the way we have designed our cuts of our crossings to get the circles in the first place, it turns out that if a circle is adjacent to another one, if they're next to each other, they must have opposite orientations. If a circle is on top of another one, if it's nested, they must have the same orientations. You can check that this is always going to be the case.

What does adding in strips give us? It gives us a consistency of colors. If something is next to 1, look what happens again in this demonstration. If this object is next to something, then when you put this strip twist, since it's next to it, it goes from blue and it switches to red, the strips behave perfectly. On the other hand, if it's on top of something, then the strip twists and it goes under perfectly so the colorings match up. Thus we see that orientation is preserved according to Seifert's algorithm.

Based on this elegant procedure how do surfaces help us better understand knots? Each of our orientable surface has 1 boundary component. If you give a knot projection into Seifert's algorithm, it spits out a surface whose boundary is the knot and we know that this surface that you have is orientable and it has 1 boundary component. What we're going to do is we're going to cap off this boundary component, and when we cap off, we glue a face onto the entire boundary. It might seem like it's going to intersect, but this is a homeomorphic gluing, this is a topological gluing.

I'm just going to cap this entire thing off. There's 1 boundary and I'm just going to cap it off, and we get, at the end of the day, an orientable surface without boundary. From our previous lecture, though, we know that surfaces have fully been classified. Thus if it's an orientable surface, it has no boundary, then genus completely determines it or in other words Euler characteristic completely determines it. That's what our previous theorem told us.

We can associate to a knot this genus of the surface. Given a knot, you feed it into Seifert's algorithm, it gives you a surface that's orientable. You cap off the surface, now you have a surface without boundary and that's orientable. It must have a genus. I'm going to take this genus and give it to my knot and now I have a knot invariant. I have something associated to my knot that's the genus of this surface that you get from building the knot. It's a beautiful idea. This is just make believe so far. This is my dream. I hope I can give this genus to this knot. I haven't shown a thing about being invariant at all. That's my dream.

Let's see what we can do. Consider the unknot. If I take a look at the picture of the unknot, here I start with a circle, I shade it in, I get a disc, now I can cap the disc off, and I get a sphere. This sphere has genus 0, so to the unknot I'm going to give it the value 0. That's good. Intuitively it makes sense. It turns out that the unknot is the only knot that you can possibly make into a sphere. Any other knot is going to be more complicated than this.

Let's take a look at the trefoil. The trefoil here, like we did earlier, will give us a genus 1, exactly like we calculated last time. It's going to be a torus with 1 boundary component. To the trefoil I can give it genus 1. I can give it the value 1 as trying to make it into an invariant.

Since we have Euler characteristic with us, there's a quick way to find the genus of a surface based on Seifert's algorithm. Here is the quick way. Instead of worrying about cutting the surface up into pieces, vertices, edges, and faces, and trying to find the Euler characteristic, there's a beautiful method, which is actually not too hard to show.

If C is the number of crossings in your knot projection and D is the number of shaded discs we get, then the genus of the surface must be given by C − D + 1 divided by 2. It's not too hard to see because the way we're constructing it using Seifert's algorithm breaks our object into pieces anyway, into these faces, and so we might as well just count the faces and the way they glue and this result quite immediately falls out.

Let's consider some examples to see what we have looked at. Look at this first example of the trefoil. We've seen this before. The number of crossings in this projection is 3 and the number of discs we have here is 2. We have 2 Seifert discs and thus the genus according to our formula is going to be 3 − 2 + 1 divided by 2, which is 1. Look at this projection here of the figure 8 knot that we built. Here it might seem the number of crossings you see is 4, so C is 4. D is 3 because you get 3 discs that we glued together in our demonstration to build our object. Thus our genus has to be 1.

What about this last object here, the trefoil, the classic projection of the trefoil? Here we see the crossing is 3, the disc, the number of discs we get is 2, and the genus is 1. In all these objects we get the genus to be 1. That's great. It looks fine so far.

When we look at these pictures, they don't look like surfaces of genus 1 again. Remember that's because of our intrinsic, extrinsic issues. Intrinsically they all are going to be surfaces of genus 1, but extrinsically we're emphasizing the boundary and not the genus.

Here again, let's look at this next example. What happens when we look at a different projection of the same knot? We've considered 2 different projections of the trefoil, but let's look at a different projection of the trefoil again. Here I take my trefoil and I'm going to do a Reidemeister III move, just pull the strand right under here. Notice that after I do this and after I take an orientation, I can cut this up into pieces by my Seifert's algorithm. To compute my genus I don't even have to finish drawing the surface.

I know what my C is, I have 7 crossings in this new projection of the trefoil. I have 4 discs from my Seifert algorithm, and so my genus has to be 2 from the formula we had earlier. But wait; we always got genus to be 1. Remember the previous 2 examples of the trefoil gave me genus 1. No matter which way I had previously drawn the trefoil, I got genus 1. Now you see that I get genus to be 2. We see that surface, that the surface that we're giving to the knot and its genus is not a knot invariant. It depends on the projection you give it.

This is sad news. A mathematician, however, tries to salvage whatever he or she can get. Remember the trick with the writhe? We perform an old trick we did earlier with the crossing number and the knotting number. Here is what the official definition of the genus of a knot is. The genus of a knot is the least genus of any orientable surface bounding the knot. No matter what surface you pick, no matter how you find a surface whose boundary is the knot—you can use Seifert's algorithm, you can come up with your own algorithm, it doesn't matter—no matter how you do this, if you find a surface whose boundary is the knot and the surface is orientable, find its genus. Do this for every possible projection of the knot and take the smallest one.

This is clearly an invariant because it's not based on the projection because you're picking the smallest one. Same reason why the crossing number was an invariant. For instance we see that the genus is 1 for the trefoil and the figure 8 because we found it to be a value 1. We know we can't get a genus any lower than 1 or else we'll end up with an unknot. So what are the properties of the genus of the surface?

We know Seifert's algorithm gives an orientable surface, we know this, we proved this. Will it always be able to give a surface of minimal genus for some projection of the knot? In other words, for any knot I want, can I find a projection of the knot such that if I put it into Seifert's algorithm, the answer it gives me is that minimal genus surface. Amazingly and unfortunately this was proven to be false by Yoav Moriah in 1987, that is there are knots for which the minimal genus surface cannot be obtained by the Seifert algorithm for any projection possible.

You need to find another way of constructing these minimal genus surfaces other than Seifert's algorithm. Thus the Seifert algorithm is great for orientable surfaces, but not necessarily for orientable surfaces of minimal genus.

Consider another powerful theorem. We have been struggling with how addition of knots and invariants are related. Remember we talked about the crossing number and how the crossing number of 2 knots added together is related to the crossing number of each knot? This is one of the great unsolved problems in math. We have a theorem, not just a conjecture about genus. The theorem says the genus of knot 1 plus the genus of knot 2 must equal the genus of knot 1 + knot 2.

We weren't able to prove this for crossing number or unknotting number, but we can actually prove this for genus. It is a beautiful proof and it's actually not too difficult to follow. This genus theorem answers a question we asked in one of our earliest lectures. We asked the question: Can 2 complicated knots be added together to form the unknot? Can you have subtractions of knots? Can they be so complicated that each 1 magically cancels out to get the unknot? By this theorem we are guaranteed that this cannot happen, since the genus of the 2 knots must be added together to get this new genus.

If knot 1 and knot 2 somehow, when you added them together, gave you the unknot, then the genus of the unknot we already know to be 0 then the genus of knot 1 and genus of knot 2 must both be 0 to satisfy the formula genus of knot 1 + genus of knot 2 = the genus of knot 1 + knot 2. If this is 0 these have to be 0. You can't have negative genus. Thus the knots, knot 1 and knot 2, must be unknots themselves.

We close with another remarkable use of genus. David Gabai was able to show that the 2 mutant knots, the Kinoshita-Terasaka knots, have different genus. This is how we were able to tell those apart. Although so many other invariants failed, genus is strong enough to succeed. We have brought the power of surfaces to knots. Although it doesn't give us quite an easy enough invariant to understand and to calculate fast, it is powerful enough to tell many knots apart even better than the Jones polynomial in some cases.

It has helped us answer some of the questions that have been challenging us. Addition makes knotting more complicated and the distinction between the classic mutant knots.

In the next lecture we still focus on surfaces, but we switch gears from using surfaces for 1-D into talking about wind flows and hurricanes on surfaces. Stay tuned.

Wind Flows and Currents

Lecture 14

Think of all possible wind flow currents of the Earth right now, flowing and changing and some stopping and moving. … [W]e ask again this question: Does every vector field on the sphere have to have a zero? Does it have to have some place where there's no wind going, or can we have it so wind is flowing everywhere all the time on the sphere?

In this lecture, we'll explore how the currents of wind flow on the Earth's surface, focusing in particular on the following question: At any point in time, is there a place on Earth where there is no wind? We begin with the study of vector fields, fields of movement. Vector fields appear in the world of analysis, which studies change.

A classic illustration of vector fields is provided by the study of a population of owls and mice. A graph shows the mouse population increasing and decreasing in relation to the owl population, with a collection of arrows representing the vector field. In the center of the field is a fixed point that represents stability between the mice and owl populations; this point is the zero of the vector field.

A natural way to get a vector field is the gradient method. The gradient method imagines that something—say, water—is flowing down from the top of the surface. The starting and stopping points of the flow are counted as zeros, places of perfect stability. A number of examples of wind or water flows on a sphere show that the flow is always continuous. Again, we ask the question: Does every vector field on the sphere have to have a zero? Note, too, that there are different kinds of zeros of vector fields, such as the center, gradient, dipole, and so on. How can we measure these zeros quantitatively?

We determine the index of a zero of a vector field as follows: We draw a small disc, a neighborhood, around the zero, then choose the top-most point of this disc and note the vector flow direction of this point. As we move around the circle clockwise in this example, the vector field rotates a full 360 degrees. In general, each time the arrow turns once clockwise, we assign it an index value of 1; each time the arrow turns once counterclockwise, we subtract 1 from the index value. The index measures how wind behaves close to the zero point.

We can also calculate the index by drawing polygons rather than circles around the regions. We place a 1 at the fixed point and a 1 at any vertex of the polygon if the vector fields point inside the polygon at that vertex. We place a –1 along any edge if the vector field points inside the polygon at that edge. We then add the numbers inside the polygon.

For any vector field on a surface *S*, the sum of the indices of all the zeros off the vector field … must be the Euler characteristic. This is a deep relationship on the way flows can move on a surface and is globally governed by the shape of the surface itself.

What happens if we add the indices of all the zeros over an entire surface? Trying several examples, we find that all spheres give an index value of 2, a torus gives 0, and a genus-2 surface gives –2. What we're really finding is the Euler characteristic. This result gives us the Poincaré-Hopf theorem: For any vector field on a surface *S*, the sum of the indices of all the zeros off the vector field must be the Euler characteristic. This shows that the way flows can move on a surface is globally governed by the shape of the surface.

The consequence of this theorem is this: If we're on a sphere with Euler characteristic 2, there must be at least one 0 in any vector field on the sphere since something must be contributing to the index. Alternatively, using our wind flow formulation, there's always a location on Earth with no wind—there has to be—because the sum of the indices must be the Euler characteristic, which is 2. ■

Suggested Reading

Richeson, *Euler's Gem*.

Questions to Consider

1. Can you draw a wind flow on a torus with no fixed points?

2. Can you draw a wind flow on a sphere with exactly seven fixed points?

Wind Flows and Currents

Lecture 14—Transcript

Welcome back and thanks for joining me again. What began as studying the shapes of the Earth has led us to the classification of all possible shapes of surfaces. This has even helped us to get a better understanding of knots, these 1-dimensional objects using 2-dimensional tools. We now turn back to the Earth again to study more about its properties. But this time our interests are in its wind flows, how the currents of wind flow on the Earth's surface.

We talk about how we can model the flow of the wind and how the shape of the Earth itself dictates the properties of how the wind flows behave. We ask the following question. It might seem like a silly question, but it's quite deep it turns out and not obvious to figure out. At any instant in time is there a place on Earth where there is no wind? Is that even possible? Must wind always be flowing at every point or can there be a place on Earth where there's no wind?

We begin with the study of something very different than what we've been attacking and looking at, which are called vector fields. So far we've only been studying static objects, things that don't move, a particular genus 2 surface, a particular trefoil. Now we're going to study movement.

Vector fields appear in the world of analysis, vector field, a field of movement. Remember how we've broken down mathematics into 3 groups: Algebra, which studies structure; geometry, which studies shapes, the world we're in; and analysis, which studies change. Vector fields belong in the world of differential equations in calculus because that's the world that measures these changes. It fits under the umbrella of analysis.

We're going to move from algebraic topology, because we've been caring about shapes in topology from an algebraic viewpoint, from a viewpoint of structures of shapes, to analytic topology, which is shapes mixed with change. We're going to switch the lenses in which we're going to look at the world. We're still going to study topology about shapes, but we're going to use a different set of tools, not the tools of structure, but the tools of change and see what happens when we consider this problem from this perspective.

Let's look at a classic example coming from differential equations to understand what vector fields are really about. Let's consider a field where owls and mice live. That's all that's happening in this field. The population of owls and the population of mice. There is a relationship between the number of mice and the number of owls, and here's the way the relationship would be modeled in its population. As the mice population increases, the owl population does as well, because there's more mice for them to eat.

As more owls are born and raised, they start eating more mice. This causes the mice population to decrease. If the mice population decreases, then there's not enough mice for the owls to eat, so the owl population starts to decrease. If the owl population starts to decrease, then now more mice can be born, so the mice population starts to increase. Visually we can take a look at this as the following diagram.

Here you see a picture of the mice population increasing and decreasing, and then increasing again related to the owl population. You see the horizontal axis is the mice and the vertical axis is the owls. Notice the owl population is again dependent on the mice population and vice versa. This is an example of the predator-prey model resulting in a vector field, these collection of arrows that tell you how the mice population and the owl population change as you follow these arrows. If mice increase then owls increase, and then if owls increase mice decrease, and on and on and on.

Notice we picked a generic population to start with and that's what gave us a circle of possibilities. If we picked a different population—a different population of mice for example—then notice what happens. The same pattern happens except now you have a bigger circle surrounding this one. The arrows are still flowing. The vector field flows in the same direction and a bigger circle would give you a different feel again surrounding this one. A smaller circle is formed with a different population near the central point.

Notice as we pick populations closer and closer to the central point called a fixed point, that there is a stability that ensues between the mice and the owl populations. The central point—this fixed point—is a place of no movement at all. Why is there no movement? The mice population and the owl population are perfectly balanced. There's no change needed because the number of mice being born is exactly compensated with how the owls are eating the mice. In the owl mice example this is where there's a stability in the system. These points are also called the zeros of the vector field.

This shows a concrete example from nature where vector fields naturally appear here. We've been looking at vector fields on the plane. As you notice this is an example of these flows on the plane, and now we move to the interplay of vector fields and shapes. Now that we have an intuition of what vector fields are about, let's see what happens when we consider vector fields on surfaces.

A natural way to get a vector field is the gradient method. The gradient method imagines that you're just flowing from the top of the surface and gravity pulls the vector field down. Let's consider examples of this on the sphere, but think of this as the flow of water on the sphere itself.

Let's take a look. Here we see a sphere with a gradient vector field on the sphere. The water is starting at the top of the sphere at north pole where you'd start pouring the water and the arrows tell you the flow, the flow of the vector field itself as it flows to the south pole. Now what if you have a torus, a genus 1 surface and I'm holding it up vertically?

If I have a torus this example's a little bit more complicated. If I pour the water on top, then the water can travel in 2 separate ways, you have a 360-degree way of movement, but there are these 2 key ways of moving where the water flows on the outside rim, goes directly to the bottom of the torus, or it can flow 90 degrees to that outside flow, it can flow to the inside genus of that torus. Once it gets there the water collects and it splits and flows through the other way of the genus all around that genus, and then it flows again from that point all the way to the end.

There are these 2 kinds of flows that are happening on this torus. There's the flow that goes right to the bottom, and these ones that kind of stop and turn directions and move the other way. If we count the number of zeros, if we count positions where water actually comes to a stop for a little bit, we see on this sphere that there are 2 zeros because there's the starting point and the ending point where the water begins to flow and the water ends the flow.

On the torus we see that there are 4 zeros. There's the very top of the torus where the flow begins from, and there's the very bottom of the torus where the flow ends in, but there are these 2 middle points where the flow comes in and stops and then switches directions, comes in and stops and then switches directions. The vector field comes in and it has to stop there to actually switch so there's a 0 there. There's some stability going on.

These are places of perfect stability, where these zeros are, where we move off of them and we are carried by the flow. Notice what happens as you come near that stability point, if you come exactly along the direction you're coming, you have to stop to change directions. But if you come not exactly at it directly head on, but a little bit to the side of it, you don't go near that 0 point at all, but you come very close, and then you turn and then you flow down.

What can we say about zeros of vector fields in general? Consider different vector fields on the sphere. These are the ones that come from the gradient flow where gravity pulls it down. Let's look at other examples. Here we see 3 different examples of vector fields on the sphere. The first 1 is that the flow is flowing left to right. It's going around the equatorial flow of this sphere. Again we see that here we have 2 fixed points, we have these 2 stability points—the north pole and the south pole—as the wind or the water flows around the sphere.

Here's another example of the sphere. Here there's that 1 central 0, 1 central 0 of the vector field or a fixed point, and the water or the wind flows around that 0 all the way until it goes through the long part from the north pole to the south pole and then it spins around this way. Notice this flow is continuous. There's no break in the flow. It beautifully flows from one to the other one. The same thing can be said about the previous example. The flow of wind and water is continuous.

Look at this third example. Here all my flows are gradient flows going from the north pole to the south pole, the water is flowing, except for this red arrow that's flowing from the south pole to the north pole. This is not a vector field because the flow here is not continuous. Things flowing in 1 direction near it have to also flow towards that direction. You can't have wind all flowing downstream and all of a sudden 1 gust of wind flowing upstream right in the middle of it without being a swirl, a change of wind in that area. All these arrows pointing down in the vector field are great as long as you don't have anything coming up immediately there. These have to be continuous flows.

These are just a few classic examples of flows on the sphere and there are numerous ones that we could come up with, and some are extremely complicated flows. Think of all possible wind flow currents of the Earth right now flowing and changing and some stopping and moving. It's quite complicated and we ask again this question: Does every vector field on the sphere have to have a 0? Does it have to have some place where there's no wind going, or can we have it so wind is flowing everywhere all the time on the sphere?

We have seen examples with places where there are 2 zeros on the sphere? We've seen this previous example where there's only 1 0 on the sphere. Can we be clever enough to have no zeros on the sphere? Is there something about the sphere that governs its properties of zeros, or are we not being clever enough? To find out we need to understand the zeros of vector fields a bit better.

First notice that there are different kinds of zeros of vector fields. Consider some classic examples. There's the center of the 0 of a vector field. It's like the center of a hurricane and the flow goes around it. It's called the center, for obvious reasons. Then there's the source where from the center point the flow goes from it. This is the north pole of the sphere and the flow flowing out of it that we talked about earlier. There's the sink where the flow comes to that fixed point. This is the south pole of the sphere under the gradient vector flow.

There are other examples like the dipole where there's a center fixed point, and the flow goes around on either direction, and there are other examples like the saddle where at the center point flows flow east and west away from it, but the north and south come right towards it. An example of a saddle visually could be something like the saddle of a horse, where you can see the water flowing towards the center of this, you could imagine the torus right here. This is the genus of the torus where the water flowing in and the moment it comes to the center it switches gears and flows east and west again. We have seen all of these on the sphere and the torus.

The dipole, for example, shows up in terms of the magnetic flow of a bar magnet. We want to measure these zeros quantitatively other than just walk through pictures. This is what mathematics does. We see something important, like these zeros, and we want to understand it and quantify it more and go behind the scenes. We did this with knots, we did it with braids, and surfaces, and now we want to do it with flows.

We define the index of a 0 of a vector field as follows. Given a source here's what we do. Let's take the source as an example that we're going to study with. We draw a small disc, a neighborhood around the 0, and what we do is we choose the top-most point of this disc, and we notice the vector flow direction of this top-most point. As we move around the circle clockwise, around this neighborhood I drew, notice how much this vector field changes when we return to where we began.

Notice at this source I'm pointing straight north. As I walk around this circle, notice how my arrows have changed. It's pointing north and then it rotates clockwise, it keeps rotating. At the south I'm pointing straight down, and when I come back again, I've rotated a full 360 degrees clockwise. Thus my index value is one because as I go clockwise once, I get a whole clockwise turn. In general each time the arrows turn once clockwise, I give my index value 1 to it. And each time the arrows turn once counterclockwise, I subtract 1 from the index value.

Let's consider some other examples. Let's look at the sink. Here my arrows are pointing in. My north part of my circle—my very top—it's pointing down. As I start going clockwise, this arrow change is actually turning clockwise again. When I'm all the way in the south, it's pointing straight up, and as I go back to the north again, I have a full clockwise turn. Around the sink, my index is also 1 because as I go around my arrows change a full 360 degrees clockwise.

What about the center? Well you can have a center where the flow flows counterclockwise, and again if I start at the north and walk around, it flows clockwise around this circle, around this neighborhood. That's a +1. What about a center that flows clockwise around it? Again if I take my north part and measure the flow direction there and walk around, notice my arrows change a full 360 degrees clockwise again. For my source, my sink, my centers in both directions, I get value of +1.

What about my dipole or my saddle? Let's take a look at those. For my dipole if I'm facing the north part of my circle, I face left, right, I'm facing west. As I start turning clockwise, my arrows turn clockwise very fast because of the dipole movement. By the time I'm at the south pole, I'm already at a full 360 degrees. If I go back again to the north pole I've completed 2 360 degree clockwise rotations. My index value is +2.

Let's look at my saddle. If I'm facing at the north of my saddle around this local area, a small region, a neighborhood around this point, I'm facing straight south. As I walk clockwise around this neighborhood my arrows start turning counterclockwise. By the time I'm at my south pole I'm facing north, but I've gone in the counterclockwise direction. By the time I come to a full 360 degree spin clockwise, my index is −1 because I've spun my dial 360 degrees counterclockwise.

These are the examples we can do to get the index of some zeros. In general, the index measures how wind behaves close to my 0 point. It's a mathematical approximation. Of course notice that my source and my sink and my centers all got value +1 so it's not a powerful measurement to tell everything apart, but at least it gives you a good measurement to differentiate those values from the dipole or the saddle.

Another way to calculate the index is to use polygons in a clever way. Instead of drawing circles around my regions I'm going to focus on drawing polygons around my region. What I want to do is around my sink, let's start with my sink, because my arrows are pointing in as an example, around my sink let me just draw a square. Instead of drawing a circle around it, I'm going to place this polygonal square around a 0. Any polygon can be used. I just chose a square as an example.

However, the one important fact is that we must make sure that all my arrows point in or out along each edge. I don't want any arrows being parallel to my edge. If you ever have any arrows pointing in or out from your 0, you want to make sure you pick a polygon so nothing points parallel to the polygon. Each edge must be measuring something going in or out, and that's easy to do. You can pick any size polygon you want so you can always do this.

For my square for example I'm going to place the 0 at the center, and notice if this is a sink, if all my arrows are going in, here's what I do. I place a 1 at the number 1 in the middle of my polygon, for my fixed point, and I place a 1 at any vertex of the polygon if the vector fields point inside the polygon at that vertex. In my square example, notice that all 4 corners of my square I have my arrows pointing in, because this is a sink. So I place +1 values at all 4 corners. I place a −1 along any edge if the vector field points inside the polygon there.

Again for my square example, all 4 edges are pointing in. Place a −1 at every edge, −1, −1, −1, and −1. What is the index? I've just done this little trick to show you what to do with the center point, you get a +1, and all the vertices and all the edges get values depending on if it's pointing in. To get the index we just add up all the numbers inside my polygon. It's that simple.

Let's add up all my numbers again. I get 1, 2, 3, 4, 5 positive values, 1, 2, 3, 4 negative values so my total index is 1, exactly what I got before. That's perfect! Let's now try it for something else. Let's try it for a saddle.

Here's my saddle. I'm going to pick a square again. It's pretty simple to do, so let's try a square and let's see what happens. I put a 1 inside my saddle at the center of the square because every value that's a 0 gets a 1 for it for free, so my inside gets a 1. My 4 corners, my 4 vertices, all my arrows notice from the flow, they're flowing out, nothing is flowing inside my square, so none of those get any values. What about my 4 edges? My top and bottom edge, those arrows are pointing in, so I get a −1 for each one of them.

On the other hand my right and left edge, those are flowing out, so I don't get anything. I have only 3 numbers in here, the −1 for the top edge, the one in the middle, and the −1 from the bottom, I add it up, my index becomes −1, exactly what I had before.

Let's try one more. Let's take a hexagon and put it around a source. I put the one in the middle because the one in the middle is always given for free for a 0 of a polygon. Every edge and every vertex all my arrows are pointing out, so nothing gets a value. Thus the only thing I have inside my polygon is that 1 that I started off with and so my index is 1, exactly what I did earlier.

Notice that the polygon method of finding the index is a discreet version of my continuous version. Remember the continuous 1 was about how my clock rotated as I walked around the circle. It's a continuous way of doing it. Discrete version means you can break it up and start counting things. For a computer scientist, somebody who wants to calculate things fast, the discrete version is much better because you can plug these finite number of values into a computer easier versus a continuous one, which might be beautiful for a mathematician to study, but might not be easy to compute for a computer.

We now show a deep relationship between zeros of vector fields and the shape of the surface the vector field flows on. We have seen several vector fields on different surfaces. We looked at vector fields, different ones on spheres, and we looked at vector fields on the torus. The vector fields, now, have a different value, they have an index associated to them. They're not just zeros there at those points. They actually have these indices we've given by these calculations.

What happens if we add up the indices of all the zeros over my entire surface? Look at this example of this vector field on my sphere. It has 2 zeros, one at the north and one at the south, but both of these 0 are a center 0. That means I give it a +1 value of my index at top, +1 value of my index at the bottom, and if I add all my indices up I get a +2 value over my entire sphere. That's great!

What about this example of a sphere? I have an index on top and an index on the bottom. This is the gradient vector field like the water flowing from top to bottom. The top point is a source. That's where my water is coming from, I get a +1. The bottom point is my sink, that gets a +1. My total value is 2 again. What about this one of the sphere where it's a dipole. Remember it just swirls around from that point all the way down going from the north to the south pole coming back up again. It only has 1 0, the dipole, but the index of the dipole is 2. My total value over all my zeros is 2 again. All my spheres have value 2.

What about my torus? Let's take my torus. Let's do the gradient flow from the top to the bottom. The top, now I have 4 zeros, so the top is a source +1, the center 1 is a saddle, remember it comes in and then flows out again. It's exactly like this example we have here. It comes in and flows out again. Here's the saddle at the top of my torus and then here's the saddle at the bottom of that genus of that torus, and then at the very bottom of this torus is a sink. You get 1 source, −1, −1 for my 2 saddles, 1 for my sink, total value is 0.

What about a genus 2 surface? If you look at the genus 2 surface and make it flow again from top to bottom, we start with a 1. We start with a −1 for that first saddle, −1 for the second saddle, −1 for the third part of the saddle, −1 for the fourth part of the saddle, and 1 for that sink at the end. Add up all these values I get −2. My spheres all give me 2. My torus gives me 0. My genus 2 surface gives me −2, glory, glory! As you can see what we're really finding is the Euler characteristic.

This beautiful structure that the Euler characteristic of the surface, if the surface's homeomorphic type, but what it's fundamentally made out of is somehow controlling the indices of my vector fields, this stunning result is called the Poincaré-Hopf theorem. For any vector field on a surface S, the sum of the indices of all the zeros off the vector field no matter what vector field you pick, the sum of the zeros of all the indices must be the Euler characteristic. This is a deep relationship on how the way flows can move on a surface and is globally governed by the shape of the surface itself.

We give a proof of this result right here live due to William Thurston. Thurston is a Fields Medal winner, one of the great ones that we've talked about earlier, and he's one of the greatest topologists and geometers of the 20th century. We will certainly talk about him more later when it comes to 3-dimensional objects. First let's actually look at his proof. The thing we do in the beginning is that given any surface—let's pretend it's a sphere for now—we partition that surface any way we want. Let's take a look.

What Thurston does is we partition the surface in any way we want as long as we observe a very simple condition. First we place a polygon, any kind of polygon you want, on each of the zeros of the vector field. Around each 0 you place a polygon, it could be a pentagon, it could be a square, it doesn't matter at all, but remember that if you have that vector field on that surface, no vector can lie parallel to an edge of the polygon.

If you're give a sphere, as our example is so far, then no matter what vector field you pick on the sphere you pick all the zeros, you place a polygon around each 1, but make sure the polygon you pick has no edges parallel to it. If there is an edge parallel to it just pick a different polygon and you'll always find 1 that works. Great that's my first thing.

The second thing is now that each one of my zeros of my vector field have a polygon around it kind of protecting it, look at the rest of the surface, and just cut it up into triangles, just partition it up. It doesn't matter how you partition it. I don't care about the rest of them. My heart is focused around the indices, so you place a polygon here and you partition the rest.

Here's what we do. We place a 1 on each vertex, a −1 1 on each edge, and a 1 in each face of my partition. All my polygons around my indices get a 1. That's great! All my triangles get a 1, all my edges everywhere get a −1 and all my vertices get a 1. This has nothing to do with the flow. This is purely on the partition. Look what happens when I sum up all these things.

If I add up all the vertices, which give a 1, all my edges, which is a −1 and all my faces which is a 1, I get the Euler characteristic of course. This is exactly how we get the Euler characteristic, and it doesn't matter how I cut the triangles up or how I partition this or which polygons I pick, by our theorems earlier the Euler characteristic does not depend on the partition. It depends on the surface itself. Now that I have my partition protecting these indices and I partition things into triangles, I place my 1s along the vertices, my −1s, along the edges, and my protected regions have a 1 at the center of those faces, and all my triangles have 1 at the center of those faces, here's what we do.

We bring in our flow. We now flow, we now place the flow on top of our surface, any flow you want. When we do this, here's what happens. The flow starts to push the values I have on my surface around. Here's what I want to do. I want to push the value 1 at each vertex—remember at each vertex there's a 1 there—my flow has to push my vertex in somewhere. My arrow has to be taking that vertex and pushing it in. I want to take my vertex that has value 1 and push it into the face the field points to. Wherever the flow points to, push it into that face again, alright.

I want to take my value −1 that's along each edge—remember nothing is parallel to it, all my edges have a flow going perpendicular to it somehow—I want to take, not parallel, it doesn't have to be perpendicular, I want to push the value −1 add an edge into the face the field points there. Consider each triangle. Notice there are 2 cases. There are 2 ways flows can come into a triangle. Since the triangles have no fixed point in them—remember all my fixed points are protected by these polygons, my triangles have no fixed point in there—so the flow that's coming in has to get out. One way that can happen is 2 flows can come into a triangle along 2 edges and can leave along 1 edge, or a flow can come in along 1 edge and leave along 2.

Remember nothing can stay in the middle so it can flow in 2 in and 1 out, or 1 in and 2 out. It's great. So what happens? If you'll look at the first case when 2 comes in and 1 goes out, then the −1 1 along this edge gets pushed in, the −1 along this edge gets pushed in, and the one at this tip vertex gets pushed in, but everything else gets pushed out. So we only have those 3 values.

I have that 1 value I've given to the center of that polygon because it's a face, that's not moving anywhere, the face sticks there, and then the −1 and the −1 from this edge and the one from this vertex, I look at the total sum, I get 0. This triangle gives me 0 value towards anything.

What about my other triangle where that 1 comes in 2 splits? The one coming in gets that −1 for that edge, but since everything else is leaving, nothing gives me anything else. Thus I have a −1 for this edge and a 1 at the center so my total is 0. Thus the sum of all values in any triangle based on the flow is 0. Each polygon which contains that fixed point that I protected is exactly the index at the 0 I calculated. Remember the indices that we talked about using the polygonal method? This is exactly how we find those indices.

What now happens? I place these numbers, the 1s, −1s, and 1s all over my polygon based on Euler characteristic. The sum of the count 1 way gives me the Euler characteristic, but the flow does not introduce new numbers, nor does it delete new numbers. It just moves it around by the flow, and all my triangles are 0 valued, but all the flow concentrates my Euler characteristic values into these polygons where the source, or the sink, or the dipole, have these attractions, but those values are exactly the indices. That's it, that's the theorem itself. The total sum in 1 count is a sum of the indices, but this is just an Euler characteristic in disguise.

We have the following immediate consequence: If we're on a sphere with Euler characteristic 2 there must be at least 1 0 in any vector field on the sphere since something must be contributing to the index. Alternatively, using our wind flow formulation, there's always a location on Earth with no wind, there has to be, because the sum of the indices must be the Euler characteristic which is 2.

Moreover if a vector field on a surface has no zeros, then the surface must be a torus. The Poincaré-Hopf theorem, however, tells us more. If there happens to be a cyclone somewhere on Earth, we see this as a fixed point of index 1. It's the center. But the Poincaré-Hopf theorem says that there must be another fixed point somewhere also because you need to add up to 2. If you have a cyclone there's something else going on that's going to give you a sum of 2. Indeed the Poincaré-Hopf theorem cares only about the homeomorphism type of the surface.

Consider this figure. Here we see examples of wind flows on the surface of a sphere or the surface of this tangled-up torus. In both cases if you count the indices of the sphere to be 1 and 1, −1 in the middle and 2 at 1 in the bottom, the sum is 2, and this complicated torus, which is just a torus tangled up, they give you 1 and 1 at the points, the top, −1 and −1 in the center, −1 and −1 at the bottom and 1 and 1 it gives you 0.

These gradient flows or any flows you can think of regardless of the shape is based on the homeomorphism type of the surface itself. This is something mathematics does. What we did was related vector fields to the zeros. Instead of trying to solve a problem about just how flows work on a sphere we generalize to a harder problem and come up with an answer that does far more.

I hope you join me next time as we push our ideas of surfaces even more in talking about curvature. See you then.

Curvature and Gauss's Geometric Gem
Lecture 15

Welcome to the greatest theorem, from my perspective, ever discovered: Gauss-Bonnet. It claims that if we know the genus of the surface, we know its total curvature.

In this lecture, we move from the rubber sheet world of topology to geometry, which also studies shapes, but it deals with rigid measurements of distances, angles, and volumes.

Curvature is one of the most visible and defining characteristics of the shape of an object. What we want to do is define curvature in a rigorous way. Consider a circle of radius r. As the radius of the circle increases, the curvature decreases; the 2 values are inversely proportional. Thus, we say that the curvature, K, at a point on a circle is to equal $1/r$. We see that the curvature of a straight line is 0, but what about a general curve in the plane?

Looking at a generic curve, we notice that curvature is defined at every point on the object. To define this, we must associate it to the curvatures of circles. Think of the curvature of a point as the best-fit circle at that point (the osculating circle). Similar to curves, we define curvature of surfaces at a point, but the process is more complicated.

The most famous definition of curvature comes from **Carl Friedrich Gauss**. We can find the Gaussian curvature as follows: We pick a point on a sphere, intersect it with a plane, and we get a circle of radius r with curvature $1/r$; then we cut the sphere 90 degrees to the original cut to get another circle of radius r, which has a curvature of $1/r$. The Gaussian curvature of the point on the sphere is $1/r \times 1/r$, or $1/r^2$.

On a surface, there exists a neighborhood around each point, giving us 360 degrees of freedom to travel from that point. We have infinitely many directions, and each one gives us a curve associated to that surface. Which direction to we pick? Gauss showed that there are 2 special directions to choose at each point on a surface. Looking at some examples, we find that the curvature of a flat piece of paper and that of a cylinder are both 0. This seems counterintuitive, but it's true from an intrinsic perspective. We see that a sphere has positive curvature, a flat sheet of paper has 0 curvature, and a saddle has negative curvature. What does that mean? The curvature shows how things bend. Negative curvature measures pulls in opposite directions and positive measures pulls in the same direction.

Curvature is one of the most visible and defining characteristics of the shape of an object.

What happens if we want to move from curvatures of surfaces at points to the entire surface? We define the total curvature of a surface as the sum of the curvature based on the surface area. For example, the total surface area of the sphere in a global setting is $4\pi r^2$. But the curvature at every point on the sphere is $1/r^2$. To compute the total curvature, we multiply the 2 values. The 2 r^2 values cancel out, and the result is a total curvature of 4π. Notice that this result is not dependent on the radius.

The Gauss-Bonnet theorem tells us that if we know the genus of the surface, we know its total curvature. This is remarkable because genus is a purely topological property, and curvature is a purely geometric property. In particular, the theorem states that the total curvature of an orientable surface, S, is 2π times the Euler characteristic of that surface. ■

Name to Know

Gauss, Carl Friedrich (1777–1855): Known as the Prince of Mathematics, Gauss is considered by many to be the greatest mathematician since antiquity. His foundational work in all areas of mathematics continues to influence our world today. We get the notion of curvature and the powerful Gauss-Bonnet theorem from him.

Suggested Reading

Cromwell, *Polyhedra.*

Devadoss and O'Rourke, *Discrete and Computational Geometry.*

Richeson, *Euler's Gem.*

Questions to Consider

1. Next time you are at a grocery store, consider the curvatures of different types of vegetables, especially kinds of lettuce.

2. Take some Play-Doh and roll it into a ball. As you deform this Play-Doh topologically, notice how the curvature changes. Do you believe the total curvature is fixed no matter how you deform it?

Curvature and Gauss's Geometric Gem

Lecture 15—Transcript

Welcome back and thanks for joining me again. Today we begin to enter the world of geometry. Thus far we have been studying objects which we can pull and stretch and bend in the rubber sheet world of topology from curves and surfaces. Although geometry also studies shapes, it deals with rigid measurements of distances, angles, and volumes.

We begin this lecture by defining what it means for curves and surfaces to have curvature, something that is fundamental to understanding shapes in nature. We then move to the greatest results in the study of shapes, from my perspective. It is called the Gauss-Bonnet Theorem. At the end of these lectures I can practically guarantee you will be shedding tears of joy. It shows that there is this amazing relationship between curvatures on surfaces given by rigidity from geometry and the rubber sheet flexibility of surfaces given by topology.

Curvature is one of the most visible and defining characteristics of the shape of an object. Remember how form and function are related in nature? Consider the curvature of objects in nature coming in different forms. We have curvatures of leaves and plants such as the numerous kinds of lettuce that you can find in grocery stores alone. Why is this so? Why is lettuce curved in that way it is?

Because the numerous creases of lettuce and even the creases in the leaves of plants allow the plants to absorb a lot of light if the creases weren't there. Curvatures of animals such as coral from the Great Barrier Reef, and seashells, show up everywhere. Of course the importance of curvature of manmade objects cannot be diminished, such as contact lenses, bullet casings, telescopes, automobile design, where aerodynamic design for the wind flow is extremely important for the curvature of the object. Even looking at simple examples like this show how drastically different some objects are between just these 2 vases and how the curvature of these objects are different.

Much less, something as complicated as this surface to see the curvature of the different points at the surface vary so drastically. What we want to do is define curvature in a rigorous way from a mathematical perspective. Our motivation for defining curvature of 2-dimensional surfaces begins just like it did for topology with understanding ideas from 1-dimensional curves.

In topology we started with knots and moved on to surfaces, and we want to do the same thing here when we enter the world of geometry. How would we define the curviness, the curvature of a circle of radius r, something that we have a good intuition for. Let's consider a circle of arbitrary radius. Let's call it radius r. As the radius of the circle increases, as we get a bigger and bigger circle, what's happening to the curviness of the object, the curvature?

Notice that the curvature starts to decrease. If you have a big circle, then the curvature is basically flattening out. As the radius increases, the curvature decreases, and similarly as the radius of my circle decreases down to a small circle, the curvature increases intensely. Thus we simply define the curvature K at a point on a circle to equal 1 divided by the radius, the inverse of the radius. They're inversely related, inversely proportional. So as the radius increases, the curvature decreases, and vice versa.

What about the curvature of a straight line, though? We want this in our heart of hearts to have curvature 0. We want this to have no curvature at all. That instinctively makes sense. This is indeed the case. For if a circle of radius r, if you keep increasing the radius of the circle so that the radius becomes infinitely large, then you see that the circle starts opening up more until the radius is infinity, and you end up with the line as a piece of this infinitely large circle. We see the curvature of this line is 1 over the radius, 1 over infinity, huge number is 0.

The curvature of this line in fact is 0. Our intuition thus far is correct. But what about a general curve in the plane? It doesn't have to be a perfect circle or a perfect line. How do we talk about curvature here?

Let's take a look and see what we get. In this example we have a generic curve, and we notice first of all that the curvature is not defined for an entire object like a line or a circle because they're so symmetric, but the curvature is defined at every point on the object. Notice for this closed curve here on the sheet of paper, at any point on this object the curvature is drastically different.

How do we define this? We associate it to something we're already comfortable with, the curvatures of circles. We think of a curvature of a point as the best-fit circle at that point. This is officially called the osculating circle. If you know something from calculus, from derivatives, you see that this is basically measuring the second derivative at this point. For example if I take this curve given here, if I pick this point here, you draw the best-fit circle at that point that approximates at that point the best circle you can get around it, and the curvature is 1 over the radius of the circle.

We generalize to define curvature of surfaces. Now that we have an idea of what curvature looks like at points on curves, what does it look like on surfaces? Similar to curves we define curvature of surfaces at a point, but how do we do this? Things get far more complicated with surfaces—as you can take a look at some of these examples here—than it is with curves.

Again, each point on a surface might have different curvature values. For example a kernel of corn, something quite simple that we see a lot, has a different curvature at the tip of that kernel than it is at the smooth end of the kernel. Even something as obvious as a kernel of corn, the curvature varies intensely.

There are several ways of generalizing from a 1-dimensional method of curvature into higher dimensions. Indeed the concept of curvature becomes powerful as we enter 3 dimensions and 4 dimension, which we discuss in later lectures, but we will stick to 2 dimensions for now.

Next we focus on the most famous and important definition of curvature, the Gaussian curvature. Karl Friedrich Gauss, 1777, was called the Prince of Mathematics. He is almost without question the greatest mathematician since antiquity, if not ever. He was the last mathematician to know every field of mathematics. In fact, due to his broad and powerful discoveries he made it so that nobody after him could possibly understand all of the mathematics and its fields he created. You can even think of him as a mini Big Bang event in mathematics he was so prolific and amazing. He revolutionized the fields of mathematics in numerous areas.

He was a child prodigy who was a perfectionist. In fact, had he published all of his notes and journals, which were not yet perfect, he might have advanced mathematics 50 years beyond what it is. The definition of Gaussian curvature of a point on a surface is given as follows: One way to try to get 2-dimensional curvature on a surface is to use our 1-dimensional intuition and knowledge that we already understand. Try to get curves on surfaces. We already know what curvature of curves look like. If I can take curves and put it on surfaces then I might be able to use my idea about curvatures on curves to get a curvature on a surface.

Let's take a look. Let's consider a sphere as our test case. What do we do to get a curve on this sphere? We can feel the curvature instinctively, but what does it mean to measure it? We can cut the sphere with a plane going through the center of the sphere through the point that we want to measure the curvature of, and when we do this, when we intersect this plane with a sphere, we actually get a curve, this great circle that goes around this sphere.

This is a circle. We know how to measure the curvature of circles. If the sphere had radius r, then the circle has also radius r, and the curvature along this circle is $1/r$. What did Gauss do with this instinct? Consider the sphere of radius r and cut the sphere to get a circle of radius r 1 way, and cut it also 90 degrees to what we just did to get another circle of radius r through this point that we picked to get 2 circles of radius r, 1 this way and 1 this way.

Multiply the 2 curvatures together to get the Gaussian curvature of the point on the sphere. Pick a point on the sphere, cut it 1 way and you get a circle of radius r with curvature 1 over r, cut it the other way 90 degrees to it to get another circle of radius r to get curvature 1 over r. The Gaussian curvature of the point on the sphere is 1 over $r \times$ 1 over r, which is 1 over R^2.

What happens for a generic point on a generic surface, a sphere is too beautiful and too perfect. Since we're on a surface, at each point there exists a neighborhood around that point, we know this is the definition of a surface, so we have 360 degrees of freedom to travel from this point. We have infinitely many directions in some sense, each 1 gives us a curve associated to that surface. If you look at my surface, I can cut it this way through that point and you get a little curve going through it. I can move around a little bit and cut it again, I get another curve, and I get another curve. I have all of these possible infinitely many curves I can get.

Gauss picked a generic direction on a sphere and 90 degrees to it, and he picked these arbitrarily in some sense because a sphere is perfect. For a generic surface, which directions out of my infinitely many do I pick? We were lucky with sphere. What do we do for a generic surface? Gauss was able to show that there are 2 very special directions at each point on a surface, no matter which surface you give him there's always 2 special directions to pick. These directions are the extrema in terms of curvature, and Gauss was able to show that these directions must always be perpendicular.

It is beyond the scope of this lecture to show why. You have to trust me on it. Each of these directions traces out a curve. I pick the 2 directions Gauss tells me to cut along 1 and the other has to be 90 degrees to it, the other. I look at the curves I get from those 2 directions, I take the curvatures of those 2 curves, and I multiply them together. We simply take the product of the 2 1-dimensional curvatures to obtain this Gaussian curvature at that point.

Let's consider some examples to see what this means in a concrete setting. Let's take a look at the curvature of a flat sheet of paper. Here, consider a flat sheet of paper. If I cut this flat sheet of paper at a point any way I want to, I get a line. If I take a 90 degrees to it and cut it again, I get another line. The curvature of a flat sheet of paper at any point I want is the curvature of those 2 lines. The curvature of those 2 lines are 0 and 0. The curvature at a point on a flat sheet of paper is 0×0, which is equal to 0.

What about the curvature of a cylinder? It turns out that the 2 natural extreme directions of the cylinder are what we feel to be the natural extreme directions, what we feel to be instinctive, which is given a cylinder, you cut 1 along the long part of the cylinder and the other 1 you cut along the circular part of the cylinder. Let's take a look.

Let's take my cylinder, pick a point—remember curvature is not defined globally it's defined at a particular point at a time—given a cylinder, I'm going to pick a point on the cylinder, now I'm going to cut my cylinder this way, this vertical cut that goes this way, this horizontal cut that goes this way. This one gives me a circle, let's say it's a circle of radius r. The curvature at this point along this circle is $1/r$. But the curvature this way gives me a line, which is 0.

If I take the curvature 1 over r and multiply it by 0 I get the Gaussian curvature at the point, which is 1 over $r \times 0$, which is 0. What Gauss says is that the Gaussian curvature of a sheet of paper is 0, and the Gaussian curvature of a cylinder is 0. The Gaussian curvature of a cylinder is 0? A cylinder isn't 0. It's not a sheet of paper, it's curved. Our instincts say something is going on. A cylinder has this curvature, Gauss what are you doing? I know you're smart and all, but somehow you're not measuring what I think you should be measuring.

Why is this happening? Again, we have problems with intrinsic versus extrinsic. From an extrinsic perspective, looking at that cylinder and that sheet of paper, of course I can see that they are different. My gut says that you should get different Gaussian curvatures. Gauss was smarter than that. He didn't want to know about the extrinsic feel of the world, he wanted to know what it looked like to live on the world. If you lived on a sheet of paper or if you lived on a cylinder near the world you live in, near your little region, your neighborhoods, it would look the same.

Why? Because I can take my sheet of paper and roll it up and become the cylinder. The distance it takes to walk to a house on a cylinder is the same as on a sheet of paper. If I take my sheet of paper, pick my point, and see how long it takes to walk to my friend's house, I could roll it up into a cylinder, that distance is exactly the same on the cylinder. This is what Gauss is measuring. He cared about intrinsic curvature, not extrinsic curvature.

Then you ask the question: Why does the sphere not have a curvature of 0? Remember we can make the sheet of paper 0, we can make the cylinder 0. How come the sphere isn't 0? Think about it. Take a sphere made out of paper. Can you make this flat? In fact to do this we need to rip the sphere apart to lay it flat, and this measurement of tearing is exactly what Gaussian curvature is doing. Since the Gaussian curvature is not 0 on the sphere, Gauss is saying you actually need to do some tearing to go from one to the other one.

In fact, this is exactly the problem with drawing projections of the Earth on flat sheets of paper. The distance in curvature causes distortions in the maps of the world we have. Gaussian curvature senses this.

What about something like a saddle? What about if we have an object which looks like this? We have curvatures going this way and a curvature going this way. Notice here that there are 2 directions we have of curvature. Let's take a look. I do the same thing I did earlier. I pick my point. Let me pick the center of my saddle, I slice it 1 way, I get a curve, and let's pretend the radius of that osculating circle is 1 over r_1. I slice it another way. Let's pretend that the osculating circle here is 1 over r_2 so I need to multiply these 2 things to get my curvature.

Notice, though, something weird is going on that choosing which direction I travel is very important because these 2 osculating circles are pulling away from each other. One circle is pointing down, but the other circle is pointing up, and because of this pull, Gauss says you need to give a direction towards your curvature. Since 1 is pointing down, let's call it a negative direction, that's $1/r_1$ negative value and the other 1 is $1/r_2$ positive value, thus when you multiply these 2 answers together the Gaussian curvature is $-1/(r_1 \times r_2)$.

Let's take a look at the sphere again, maybe we missed these orientation issues. If we take a look at the sphere, let's pretend we want to orient the down direction as negative. Let's just pick our direction just like the orientation of a knot, we can just orient this down direction of this sphere as negative. Then if I take my 1 slice, this direction will be $-1/r$ and this direction would be $-1/r$ as well.

If I multiply this direction by this direction I get $-1/r \times -1/r$, which is positive $1/r^2$. In this sphere if I point this direction to be positive or this direction to be positive, at the end of the day my Gaussian curvature doesn't change. The most important thing is here for this saddle point, if I point up positive or down positive, the fact that they have opposite poles is what that negative sign is about. It's not the orientation you pick, just like knots, it doesn't matter which way you orient it, you're going to end up with the same object at the end of the day anyway, same thing for surfaces for Gaussian curvature.

What can we say about positive 0 and negative curvatures? Now that we have a sphere being positive, a flat sheet of paper or a cylinder being 0, and the saddle being negative, what does that mean? They show how things bend and this extrinsic phenomena, this negativeness is an extrinsic data where the negative means you're pulling against each other and positive means you're going towards. Zero is flat, positive measures again, pulls in the same direction, and negative pulls in the opposite direction.

They also show from an intrinsic perspective the value of land. You see for a 0 curvature, in some sense, the world you can think of where we live in, the city you live in, in some sense is flat. The Earth we live in has curvature, but locally it looks flat, so let's pretend for 0 curvature an area of land is $10. Imagine you live in a flat plane, area of land is $10. If you live in a world of positive curvature, the value will increase. The Earth itself, although you can say it's $10 per 1 part, the whole Earth has a positive curvature. The value will increase because land becomes rare, which it is for the Earth.

That's why property values always go up, because you're running out of space because the world is curving on itself. For negative curvature, the value will decrease, it will be less than $10 an acre. For positive curvature, it will be more than $10 and for negative it will be less, because there's going to be so much land around you because of the negativeness.

Notice this is an intrinsic phenomenon. It happens because you live on the world, not the way the world itself looks. For example, the curvature of leaves, lettuce, and plants in general are mostly negative curvature because they want a lot of surface area. They try to pack in as much surface area as they can to absorb light.

Thus far we've been concerned with curvatures of surfaces at points. What happens when we want to study the entire surface as a whole? Let's consider the sum of the curvature of every point on the surface. There are clearly infinitely many points on the sphere so you might say the sum just doesn't make any sense. You're going to get infinity as your value. This does not necessarily have to be the case if we think about it in a different way.

We're going to define the total curvature of a surface as the sum of the curvature based on the surface area. Let's consider the sphere as an example. The total surface area of the sphere in a global setting is $4\pi r^2$ if the radius of the sphere is r. But the curvature at every point on the sphere is 1 over r^2. We measured this. We actually computed this, so what does this mean. It means let's compute the total curvature, now that we have the curvature at each point and we know how much stuff we have, we have $4\pi r^2$s worth of it, then we just multiply the 2 values.

The total curvature of a sphere is $4\pi r^2$s worth of material $\times$ $1/r^2$ at each point. If you multiply those 2 values together the r^2s cancel out. You get 4πs worth of total curvature. What I want us to notice is that the 4πs worth is not dependent on the radius itself. So no matter the size of the sphere, the total curvature is always 4π, this is amazing. You pick a small sphere, the total curvature is going to be 4π. Why is that? Because a small sphere has a lot of high curvature, but the surface area isn't that much. A big sphere doesn't have that much curvature at all at every point, but you add it all up, and it works out perfectly. This seems like a very, very nice coincidence since the sphere is a very, very nice object.

Let's look at the total curvature of a torus. Let's see what we get here. It turns out that the total curvature of a torus, if you add it up—now this is hard to compute because you do need geometry to do this, you need differential geometry to understand, but it turns out that the total curvature of the torus, if you add up all the values—is going to be 0.

In other words, this curvature on the outside part of the torus—notice that it's positive, this thing shaded blue in your diagram—it's positive because the pull is both facing in the same direction. You have a curve going this way and a curve going this way. The 2 90-degree curves that give you the Gaussian curvature are pointing in the same direction so it's a positive curvature like the sphere, but the stuff inside are all looking like saddles.

There's 1 going this way and the other pulling the opposite direction, and if you put the torus right on a flat table, you see that the top rim and the bottom rim the curvatures are exactly 0. They're flat, you can lie them flat on the table.

You see the curvatures range from this positive towards a 0, towards a negative, back towards a 0, back to positive. What happens when we add it all up? It goes away to 0. The amount of positive cancels out perfectly with a negative regardless of the size of the torus.

What about these other shapes? What do we get here? Look at this shape. It's a sphere with a little divot. The sphere has total coverage for 4π, but you've increased the curvature at this little tip because you've pushed it up and it's really sharp now. You've increased curvature, but right around that divot to make the divot, you had to introduce a negative curvature right around it.

What about these other curvatures, this torus pulled this way or this knotted-up triple genus torus? Welcome to the greatest theorem, from my perspective, ever discovered Gauss-Bonnet. It claims that if we know the genus of the surface, we know its total curvature. The genus tells us the total curvature? Genus is a purely topological property. It's not a geometric property. The genus is just rubber sheet. The genus has nothing to do with geometry and curviness.

Curvature is a purely geometric property. In particular, the theorem states that the total curvature of an orientable surface—all the ones we're looking at, surface S—is basically 2π times the Euler characteristic of that surface. If you know the Euler characteristic you multiply it by 2π you get the complete total curvature. The curvature of the sphere is $2\pi \times$ the Euler characteristic is 2, 2π times 2 is 4π just like we got.

The Euler characteristic of this donut, of this torus, is 2π times the Euler characteristic is a 0, which is equal to 0. Notice this is not a geometric sphere in this example. Look at this example again. This sphere is not a geometric sphere, but it's a topological one. Gauss-Bonnet says the total curvature of any sphere, which is homeomorphic to the perfect sphere of any sphere, has to be 4π, which means the total curvature of this sphere with the little dimple, the little divot, is also 4π no matter how you deform it, because the negativeness around this little dimple cancels out with an extra positive curvature you introduced by making it sharper perfectly, perfectly. That's what Gauss-Bonnet says.

In fact look at the curvature, the total curvature, of a torus that looks like this. I know the total curvature is going to be 0. In other words all the negativeness that you see here in red and all the positiveness that you see in other places cancel out perfectly. If I stretch it even more, it doesn't matter. The amount of stretching will introduce just enough negative and positive curvatures to cancel out beautifully. The left side of the Gauss-Bonnet Theorem, the total curvature is purely geometric constraint and the right side is a purely topological one.

Thus this surface of genus 3 its total curvature is $2\pi \times$ the Euler characteristic which is -4, $2\pi \times -4$ is -8π. I know its total curvature without even having to compute a thing. Geometry, the curvature at every point, and topology, this global phenomena of genus are intrinsically related. What a gorgeous, gorgeous result!

I want to close this lecture by considering how curvature and Gauss-Bonnet behave, not on beautifully smooth surfaces, but on polyhedra. We've been looking at polyhedra motivated by the platonic solids. Let's look at polyhedra in a different way based on Gauss-Bonnet.

Look at this polyhedron here. Notice that curvatures at points of the faces of the polyhedra at any point in the face is going to be 0 because it's flat. Also curvatures along edges, any point along this edge, it's going to be 0 also because if I take an edge of a polyhedron, I can just intrinsically go like this, the way I live on this world hasn't changed at all if I go like this or this, so the curvature is this side is 0 curvature, this could have some curvature here, but it's 0 times whatever and it's going to be 0.

What does that mean? That means for polyhedra all my curvatures concentrated at the vertices. For something like this all the curvatures here are 0 at faces. They're 0 along edges, but they're all concentrated at vertices.

How do we calculate the Gaussian curvature at a vertex. At a vertex the curvature is 2π minus the total face angles at that point.

Let me give you an example. What is the curvature of this point in this icosahedrons? The curvature of that point of the icosahedron is just cut it open and lay it flat. It's 360 degrees, it's 2π minus the amount of angles you have here. In other words the total curvature is the angle left over when you lay it flat. The total curvature of this is just this angle. Notice what happens—this is a pretty curvy point—so you can see how much curvature it is, but what happens if I take off one of these?

The point gets more curved. It gets curvier, and look, the angle increases. Let's do it again. Look, the curvature it gets sharper at this point and the angle increases. The curvature at this point is π. Why is the curvature at this point π? Because the angle here is $2\pi - 180$ degrees, which is π, so this angle right here is π and that's the curvature.

The polyhedral Gauss-Bonnet says the following thing: The sum of curvature is $2\pi \times$ the Euler characteristic of the polyhedra. That's exactly what it says. Consider the examples of a cube and a tetrahedron. Here, take a look at the cube. We only need to worry about the vertices of the cube because that's where all the curvature is, but computing the curvature of the vertex right here is $2\pi -$ the face angles. You have 3 angles meeting at that point, each 1 is 90 degrees. It's $2\pi - 3\pi$ over 2, which is simply π over 2.

The curvature at the corner of a cube is π over 2, but how many corners does this cube have. It has 8 corners. The total curvature is $8 \times \pi$ over 2, which is 4π. That's exactly what we got for the sphere. The sphere was 4π and this is 4π and that makes sense because this is a topological property. Locally at every point it's geometric, but globally it's a topological one.

Consider this example of a tetrahedron. At every point it's 2π minus, you have 3 of these angles meeting, $3 \times \pi$ over 3 angles. The curvature at this point is π. The total curvature is 4 corners, $4 \times \pi$, which is 4π just like before.

Let's consider a more complicated polyhedron like this torus. Look at the corner of this torus. There's these outer 4 corners whose curvature is $2\pi - 3$

$\times\ \pi$ over 2 because you have π over 2 information there, which simplifies to a total of π over 2 and then you have these inside corners. You have 8 inside ones. Those inside corners look like this. It has these 3 squares here along with 2 more squares, a total curvature of $-\pi$ over 2. You have π over 2, π over 2, π over 2, π over 2, π over 2, so it's $2\pi - 5\pi$ over 2, total curvature is $-\pi$ over 2 which makes sense, because the inside of that torus has that saddle-like feel that we have.

We see that the inside of that torus is 8 vertices of $-\pi$ over 2. The outside corners of the torus are 8 vertices of π over 2, you add it all up you get 0. In fact, we can do something far more complicated with this example which we have seen earlier, which we computed the genus for. Here I have curvature at points like this, which you get π over 2, which you can calculate. Points like here, which you get $-\pi$ over 2, which you can calculate, again and other kinds of points like the inside of this one which you get $-\pi$ over 2 also.

You add up the total curvatures of the number of these different kinds of points. You do all of the work of algebra and you get -4π. If it's -4π now I can reverse engineer it, -4π which means my Euler characteristic must be -2 because $-2 \times 2\pi$ is -4π, which means my genus must be 2. What a beautiful thing. I can find the genus based on Gauss-Bonnet reverse engineering it.

What have we learned in this lecture? We have seen how to understand 1-dimensional curvature and used it to define 2-dimensional Gaussian curvature. We have seen different kinds of curvature, positive 0 and negative and its implications to intrinsic and extrinsic viewpoints, and we have tied into topology of surfaces to the geometry of curvature by the fabulous Gauss-Bonnet Theorem.

In our next lecture we push our idea of geometry even further by coming to grips with the most famous geometric value, area. Stay tuned.

Playing with Scissors and Polygons
Lecture 16

The Bolyai-Gerwien theorem states the following thing: Any 2 polygons of the same area are scissors-congruent. Can you imagine any 2 polygons, no matter how crazy they are, as long as they have the same area, you can take a pair of scissors, cut these into finite pieces, rearrange it, and get the other one? This means that the only quantity to measure scissors-congruence is area.

In this lecture, we move into discrete geometry, which provides us with a world of approximations that we can use to study the natural world. We've already seen that polyhedra approximate spheres. A deep theorem in mathematics says that any surface, no matter how complicated, can always be approximated by a discrete surface.

We begin with a new notion of equivalence of polygons: Two polygons, P and Q, are said to be **scissors-congruent** if P can be cut into smaller polygons such that those pieces can be rearranged to form Q. We see a square and a triangle and a square and a cross that are scissors-congruent but are not congruent in the geometric sense. Note that the shapes must have the same area. Scissors-congruence is a weaker form of geometric equivalence, just as homeomorphism was a weaker form of isotopy.

If 2 polygons have the same area, can we always make them scissors-congruent? If not, then what characteristics are we looking for to find scissors-congruency—angle, side lengths? We'll try to build some tools to answer these questions.

We first see that any triangle of some area can be made into a rectangle of the same area using a finite number of cuts. Using a more complicated procedure, we also see that any rectangle of some height and some area can be made into any other rectangle of a different height and the same area. We also need to note here that any polygon can be triangulated. Given any polygon, we can cut it into triangles by drawing a diagonal from one vertex to another.

So far, we've built the following tools: (1) any triangle can be cut up by scissors-congruence to make some rectangle, (2) any rectangle can be cut up by scissors-congruence to make a different rectangle, and (3) any polygon can be triangulated. These tools allow us to prove the Bolyai-Gerwien theorem, which states: Any 2 polygons of the same area are scissors-congruent.

Given 2 polygons, *P* and *Q*, we cut each polygon into triangles. Next, we convert each of the triangles into some rectangle. Then, we convert each of the rectangles into the rectangle of choice as long as it has the same area and base length 1. We stack up all these rectangles to get a super-rectangle of base length 1, and the height of this super-rectangle is exactly the area of the polygon because area is base times height.

We say 2 polygons, *P* and *Q*, are scissors-congruent—not congruent but scissors-congruent—if *P* can be cut up into smaller polygons where these pieces can be rearranged to form the polygon *Q*.

Can we apply scissors-congruence to 3-dimensional polyhedra of the same volume? Unlike polygons, which have one notion of angle, polyhedra have 2 notions of angles. A polygon has only the corner angles, but polyhedra have face angles and dihedral angles (the angle formed between 2 faces). The solution to this question can be roughly stated as follows: If 2 polyhedra, *P* and *Q*, have different kinds of dihedral angles, then they cannot be scissors-congruent. In other words, volume is not enough. We show 2 tetrahedra that have the same base and height but are not the same because they have different kinds of dihedral angles. We could never cut one up and rearrange it to make the other. ■

Important Term

scissors-congruent: A notion of equivalence. Two objects are scissors-congruent if one can be cut up and rearranged into the other.

Suggested Reading

Devadoss and O'Rourke, *Discrete and Computational Geometry*.

Questions to Consider

1. Cut two rectangles of the same area from paper. Try to cut one rectangle and rearrange the pieces to form the other.

2. Can you think of ways to find the least number of cuts needed to make one rectangle scissors-congruent to another?

Playing with Scissors and Polygons

Lecture 16—Transcript

Welcome back and thanks for joining me again. The natural world around us is extremely complicated, from leaf formations to protein folding, to DNA entanglements, to quantum fields and physics. One way to get a grasp on perplexing shapes and designs is to approximate them using simpler shapes. How can we try to approximate a leaf? How can we try to approximate a tree? If we can fully try to understand the shape of a tree itself, we need intensely powerful tools, layer upon layer of data.

Most of the time we've been studying surfaces that are smooth and continuous. This is how we came up with Gauss-Bonnet's ideas. Now we move into discrete geometry. What is the purpose of discrete geometry? First it provides us with a world of approximations. We've already seen this in the world of polyhedra, which approximates spheres. Polyhedra, or the platonic solids as specific examples, are actually isotopic to spheres, but from a geometric perspective—which is the lens we want to look at today, in terms of area, volume, and length—polyhedra aren't exactly perfect spheres, but they come pretty close to making things look spherical. A deep theorem in mathematics says that any surface, no matter how complicated, no matter how high the genus is, can always be approximated by a discrete surface. In other words, no matter how complicated your surface is you can come as close to that surface as you want to, if you pick a discrete polyhedral surface small enough, with small enough triangles and small enough faces, to come close to that surface. This approximation is fundamental in the world of computer science.

Calculations for example can be done much more easily in a discrete setting where you're worried about polygons and polyhedra and vertices, edges, and faces than in a smooth continuous setting, since this is how computers accept and manipulate data. If you try to feed a smooth surface into a computer data system, it's hard for it to understand and easily accept what it is. If you give it data based on points and lines and triangles and pentagons glued in certain ways, however, it's far easier for the computer to do this. A computer is designed to handle points, lines, polygons, than smooth surfaces where curvature varies continuously.

The influence of discrete geometry ranges in nature from protein modeling to even ideas in facial recognition software, surface reconstruction, robot motion planning, and so much more. We even considered the discrete setting the last time in our previous lecture when we closed with looking at the Gauss-Bonnet theorem from the perspective of polyhedra.

Thus our lecture today will focus on simple notions of just polygons, the building blocks of discrete geometry. You might be thinking we've already studied polyhedra, these surface-like approximations, why go back to polygons, something I've understood since I was a kid. What does it mean for 2 polygons to be equal? From a geometers perspective, 2 polygons must not only have the same area, but the same features as well. For example, 2 polygons must have the same angles. They must have the same side lengths and so on and so forth. In other words, they must be congruent, they must be perfectly identical.

We now define a new notion of equivalence of polygons and here's what we say. We say 2 polygons, P and Q, are scissors-congruent, not congruent, but scissors-congruent if P can be cut up into smaller polygons where these pieces can be rearranged to form the polygon Q. I'll say that again. Two polygons P and Q are scissors-congruent, or equal, this new concept of equivalence if 1 can be cut up into smaller pieces where these pieces can then be rearranged and be made into the other polygon. What does it mean for 2 polygons to be scissor congruent in a concrete setting?

Which polygons do we know are scissors-congruent to one another? Let's take a look. Let's actually try to show scissors-congruent for some examples. Here we start with a triangle and this is 90 degrees at this corner, 2 45 degrees at this corner—and again this is just an approximation for a demonstration—but I want to know whether this triangle is scissors-congruent to some square. Is there a square out there that's going to be scissors-congruent to this triangle? Notice the notion of scissors-congruence here. What I want to show is to use a pair of scissors to cut this triangle into pieces, rearrange the pieces to get that square.

Let's try it. I'm going to cut it straight down the middle, draw this perpendicular bisector straight down, and I'm just using a pair of scissors and straight line cuts, and now I can rearrange the pieces here, and you can form a square. Notice this square and this triangle must have the same area because as I make these cuts I cannot throw anyway any pieces, I must use all the pieces of the puzzle to reconstruct the other one. If I had a particular area before, I'm going to end up with the same area again.

What about some shape like this? Is this shape scissors-congruent to a square? Let's pick a square as an example. Is there a square of the same area as this region, this red region, that's scissors-congruent to it? Let's try it. Let's see if I can pull this off. What if I start cutting here, cut off a little triangle at that corner, and I'm going to cut off another triangle at this corner, third triangle at this corner, and a fourth triangle at this corner. It doesn't look like much of a square so far, but as you probably have guessed I probably have practiced this before.

If we can fit each of those triangles in these corners the way I've cut them, all I'm doing is I'm cutting this piece and I'm just swinging it over, and I'm cutting this piece and I'm just swinging it over, you get a square. For these 2 particular examples, I can take that triangle and make it into a square, I can take this cross-shaped object and make it into a square. Notice that in both of these cases the 2 polygons are not congruent. The square was not congruent to the triangle in the geometric sense, and the square was not congruent to the cross in the geometric sense. This notion of equivalence, 2 polygons being equivalent in this way, is a weaker notion than congruence.

Scissors-congruence is a weaker form of geometric equivalence just like homeomorphism was a weaker form of isotopy when we're comparing surfaces. Notice that we only allow a finite number of straight cuts. If you allow me to have infinitely many cuts than I might as well use integration from calculus. Remember my goal is to think of it from a computer science perspective. What would a computer accept as data? It can accept as many cuts as you give it, but you can't say cut it forever. It doesn't know what to do with that concept of forever.

We can cut it a million times, 20 billion times, but I cannot say keep doing this forever. I need to have a finite number of cuts to work with. I also assume that no area is lost from the cutting. Remember whatever I started with before, at the end of the day after I cut and rearrange, I must end with the same area at the end of the day. I cannot throw pieces away. Given 2 polygons how do we know whether or not they're scissors-congruent?

Of course if 2 polygons have different area, they cannot be scissors-congruent. There's no way I can take a polygon of area 1 cut it up and make it into area 2. I can't make area appear and I can't throw away pieces. We know they must have the same area to be possible candidates for scissors-congruence, but if they have the same area, can you always make the 2 polygons scissors-congruent? Is area that powerful of a measure? If not, if 2 polygons have the same area but they're not scissors-congruent, then what characteristics are we looking for? Is it angle, side lengths?

Let's take a look at these examples. Here we see 3 polygons, all of equal area. One looks very spherical in nature, the other it looks like a jagged comb, and the third looks like a piece of a block made out of 90-degree corners. Can I make all of these into each other? Since they all have the same area, is there a way I can take that almost spherical polygon, the first 1, cut it into pieces and make it into the comb? Can I fit my polygon, the spherical 1, into small enough shreds so they fit into the tips of the comb and yet have exactly the right pieces left over remaining to lock the rest of it out, exactly fitting to get the second one and to get the third one. This simple problem gets to the heart of discrete geometry. It forces us to find out what polygons are really about. Is it really about area or is something else going on?

We will try to build some tools we can use to answer this problem 1 way or the other. First let's consider the simplest polygons to build some simple intuitive reasoning. Our first result will show that any triangle can be made into some rectangle. Consider any triangle you want, the first thing I want to do is I want to place the longest side as its base on the floor. Once I have the longest side on the floor, I'm going to take a horizontal cut exactly halfway up this triangle. I have a triangular top and then I have this quadrilateral that it's sitting on top of. Remember I can have as many cuts as I want, but it has to be a finite number of them.

I take that horizontal cut and then I draw a perpendicular bisector straight down, so a plumb line from the very tip all the way to this halfway point.

That cuts my top triangle into 2 little triangles. I just swing those 2 little triangles around and I get this perfect rectangle. It's the same kind of swinging I did to go from that cross example to that square. This shows, given any triangle you want, I can make it into a rectangle using a finite number of cuts, that any triangle and some rectangle are scissors-congruent as long as they have the same area. I can take this triangle of area 1 and make it into some rectangle again of area 1.

My second result says that any rectangle you give me can be made into any other rectangle I want of the same area. First we show that a triangle can be made into a rectangle. I'm going to show that your rectangle can be made into my rectangle as long as they're the same area. Here's the setup. Here we have 2 rectangles, rectangle 1 and rectangle 2, and rectangle 1 let's say has height 1 and length 1 and rectangle 2 say has height 2, length 2. We're going to assume that the height of 1 and the height of the other actually differ because if they were the same, then they must have the same length, they must have the same area, and so there's nothing else to talk about, they're already congruent. They don't even need to worry about scissors-congruent.

I'm going to assume that the height of the second 1, H_2, is going to be smaller than the height of the first one. The first one is taller, but remember they have the same area. That's what we're assuming in order to show scissors-congruence. Thus the length of the first one has to be smaller than the length of the second one. So we assume H_2 is less than H_1, which could be less than or equal to L_1, which is going to be less than L_2. Then what do we do? We want to be a little careful about the second rectangle that it's not too long and skinny.

Here, since length 1 is less than length 2, then R_2 is longer than R_1, but we don't want R_2 to be too long. We don't want it to be extremely long and skinny. What I do is, if the length of the second 1 is bigger than twice the length of the first 1, the second 1 is bigger than twice its length if it's so long, I'm going to cut the second in half and I'm going to take that 2 pieces and stack 1 on top of the other one. I have a rectangle half the length, and if it's still too long I cut it again and I stack it, I cut it again, I keep doing the stacking formation until the length of the second 1 isn't too long and skinny.

At the end of the day by doing this procedure I get H_2 is less than H_1, it's less than or equal to L_1, which is less than L_2, exactly like before, but now I'm going to make another condition that this last L_2 is less than 2 times L_1, that the length of the second 1 isn't too long. I'm going to chop it and stack it up till it's not too long. Once I do this, look at what I need to do next.

I put my 2 rectangles and I stack them flush against each other, and I take a slice from the northwest corner of the one rectangle to the southeast corner of the other rectangle and I draw this diagonal slice. When I do this, this cuts my rectangles, my 2 of them that I stacked flush against each other, into pieces.

The reason I'm able to make this cut and have this cut stay within the 2 rectangular regions is because the second rectangle isn't too long. If it was extremely long, than this slice would not work. I push the second rectangle, keep doing the stacking trick until I have this rig set up so when I do the slice, I get the following pieces. Rectangle 1, according to this slice, cuts into 3 pieces. You have this pentagon, which I'm going to call C, and you have 2 triangles A_1 and B_1, at the same time rectangle 2 gets 3 pieces from these cuts. I get a pentagon I'm going to call C again, and I'm going to get 2 triangles, A_2 and B_2.

Notice that these pentagons, these Cs from rectangle 1 and rectangle 2 are identical. They're both the same piece because they're overlapped right on top of one another. What I'm going to show you is that all these pieces are the same. In other words the C from one and C from the other are the same, which we already know, but I'm also going to show you that A_1 and A_2 are the same piece, and B_1 and B_2 are the same piece, and if I can do this then I've basically convinced you that the rectangle 1 can be made into rectangle 2 with just these cuts and rectangle 2 is just arbitrarily given to be the same area as rectangle 1 but any length and width you wanted, as long as the areas were the same.

How do we do this? How do we share that area 1 and area 2, those 2 triangles are identical. If we consider the triangle A_1 and A_2, and the bigger triangle T as shown here, we see that A_1 is similar to T, they're similar triangles. They have the same scaling. In the same way A_2 is also similar to T. Let's look at

A_1 in detail. A_1 has height $H_1 - H_2$, the height of the big rectangle minus the height of the little rectangle in terms of heights, but what is the length of A_1, that triangle? I don't know. But the height of T is H_1 and the length of T is L_2, it's the longer one. Remember T is that big triangle we get.

Thus, since A_1 and T are similar, I can do the following thing. H_1 minus H_2 is to H_1 as X, the length of this A_1 piece which I don't know, is to L_2, there's this ratio because they're similar triangles. Thus if I solve for X, to solve for the length of A_1, I use some simple algebra, I cross multiply, and then I notice this one thing. L_2 times H_2 is actually the same thing as L_1 times H_1 because they're both the area of those rectangles and I know my rectangles were the same area. I can use this substitution, once I cross multiply and once I factor things out, I substitute in, and then I recombine like terms and notice my H_1s cancel and I'm left with my value of X.

The length of X which is L_2 minus L_1 is exactly measuring the length of that small triangle. But this length of that small triangle is exactly the length of A_2. It turns out that A_1 has height H_1 minus H_2 in length, L_2 minus L_1, but A_2 has length L_2 minus L_1. But I know that A_1 and A_2 are both similar, they're both similar to T, which means they're similar to each other, which means they're the same. I have shown that those pentagons, the Cs, were identical and now I've just shown that the A triangles, A_1 and A_2 are identical. Using the same argument I just did, you can show that B_1 and B_2 are identical. They're congruent. Thus we have proved that any rectangle can be made into any other rectangle using simple scissors-congruence.

We're going to pause and prove a statement we have used several times in other lectures, something I've kind of swept under the rug, and the statement is: Any polygon can be triangulated. Given any polygon, I can cut it into triangles by taking my pencil and putting it at one of the corners of the polygon and drawing a line to the other corners of the polygon. This is what a triangulation means. It's a diagonal to go from one vertex to another vertex which breaks the polygon into triangles. We use this result for example for Euler's formula for polyhedra. We needed to cut it into pieces and show that every edge I draw cuts the face into 2 pieces and it increases an edge. We used this many times before. There are 2 main reasons I want to prove this. Although it might seem obvious it takes some work to actually show, and secondly the way we prove it uses visual ideas motivated by computational geometry, a bit different than just theoretical mathematics.

I want us to learn this new way of thinking about stuff. Consider the following polygon: If I can find just 1 diagonal for this polygon, just 1 diagonal that goes from one vertex to another vertex that stays inside the polygon, I claim I'm done. If I find 1, I claim I'm done. Why is that? Because if you can always find 1, here's what you do. You take the polygon, you draw that 1, it shatters into 2 smaller polygons. You just told me given any polygon you can always find one. That means each one of the smaller polygons must have a diagonal in there.

Great I draw 1 for each one of those. That shatters those into smaller polygons. But you told me each polygon that you give me there's always 1 diagonal. I take a diagonal, I throw it in there, for each one of those it keeps shattering. I do this over and over again and I'm left with triangles. It's the smallest piece you can get. If I can always show that every polygon in the world has at least 1 diagonal, I can just use this method to show I can always triangulate things. How do we know there's always 1 diagonal for every polygon in the world? I just need to find 1 and I've automatically gotten an entire triangulations worth of polygons. How do I find that 1?

Look at the beauty of this proof. It's very different than the mathematical proofs we've thinking. It's the way a computer scientist would think. Let's take a look. Let V be the smallest point of my polygon P, P being the lowest point you can find. Since it's a point on my polygon it has 2 vertices on either side of V because it's part of a polygon. Let A and B be the 2 vertices on either side that's adjacent to V. Draw a line from A to B. If this line is a diagonal, I'm done. That's it, I've found the one diagonal.

What if this line isn't a diagonal? In other words, what if some part of the polygon wraps back inside and comes back out so when I draw this line it leaves the polygon, which does not make it a diagonal? If it leaves the polygon, here's what I do. I take that line AB and I just move it down, keep it parallel all the way down and I place it at V. I start it at V and I slowly sweep up, do the sweeping method from V all the way towards my AB. And as I sweep up I must hit some point called X first. There must be a vertex I hit first. If I don't hit anything and if I go all the way to AB, then I have my diagonal. If I hit something, however, then that point X, you can draw a line from X to V, and there's my diagonal. That's perfect!

What have we shown? We've shown every triangle can be cut up by scissors-congruence and made it in to some rectangle. We've shown that any rectangle you give me by scissors-congruence, can be made up into any rectangle I want by the fact that they must have the same area, and finally we have shown that any polygon can be triangulated. It turns out that we have all the tools we need to prove a stunning theorem.

The Bolyai-Gerwien Theorem states the following thing: Any 2 polygons of the same area are scissors-congruent. Can you imagine any 2 polygons, no matter how crazy they are, as long as they have the same area, you can take a pair of scissors, cut these into finite pieces, rearrange it, and get the other one. This means that the only quantity to measure scissors-congruence is area. We don't need to worry about length or angles or distance, area is enough. Thus, all these crazy shapes we talked about earlier are scissors-congruent.

The proof of this theorem is as follows: Given your polygon P and my polygon Q, here's what I do. First I cut each polygon into triangles. Can I do this? Absolutely, I have the power of cutting each polygon into triangles. That's great! I take each one of my triangles and I convert it into some rectangle. Can I do this? Absolutely I can. I can take my polygon first, I shatter it into triangles, convert each triangle into some rectangle for your polygon P and my polygon Q, and then I take each one of these rectangles and I'm going to convert it into the rectangle of my choice as long as it has the same area.

I'm going to convert all of your rectangles into the rectangle I want, but my rectangle must have base length 1. I don't care about the height. All my rectangles have base length 1 and its height is whatever you need it to be so the area works out. Now what do I have? I've taken my polygon, shattered it into triangles, converted each triangle into some rectangle, converted those rectangles into the rectangles of base length 1. I'm going to stack up all these rectangles to get a super rectangle of base length 1, and the height of this super rectangle is exactly the area of the polygon because it's base time height.

Remember we haven't lost any area. Thus I've converted your polygon into this rectangle base 1, height, area, and my polygon I can do the same procedure of taking my polygon and making it into a rectangle of base 1 and height, the area. I've made your polygon into the same object as my polygon and thus I can make yours into mine.

This proof of the Bolyai-Gerwien Theorem is not just theoretical, it's not just a there-exists-a-method-to-do-it way, but it's a practical constructive proof. You can build it today. You can actually go and test this procedure on any polygon you want. It is a proof we built with our own hands. This is a very rare mathematical phenomena where we are usually just interested in true/false statements. Can we or can't we do it? However this constructive proof we did today, although a little different than previously, is highly useful and needed in computer science.

I want to close this lecture by extending this idea of scissors-congruence into 3 dimensions. The question is: Can I take a 3-dimensional object, one of volume 1 and another 3-dimensional object of a polyhedron, another of the same volume, volume 1, and if these 2 3-d polyhedra are of the same volume, can I cut this with a saw and rearrange the pieces and make it into this one? We were able to do it for polygons and area, does this work for polyhedra and volume with blocks of wood? Can I cut a block of wood and rearrange the pieces and make it into the other 1?

Probably the most famous speech in all of mathematics, Professor David Hilbert in 1900, spoke to the International Congress of Mathematicians in Paris. Here he announced what he believed to be the 23 problems which would define mathematics for the next 100 years. The third of his 23 problems stated the following: Is it possible that 2 tetrahedra with the same volume not be scissors-congruent? Is it possible to have 2 tetrahedra of the same volume? He didn't worry about polyhedra in general—2 polyhedra being volume scissors-congruent—he just said can you have 2 tetrahedra be scissors-congruent or not? In other words he asked the question: Is volume the key ingredient like area was for polygons? Unlike polygons, which have

1 notion of angle, polyhedra turn out to have 2 notions of angles. For a polygon, there's only the corner angles that we talk about, but for polyhedra we have these other angles. There are 2 1s, so the first 1 called the face angles of the polygons make up the polyhedra.

Given a polyhedra, like this cube, we have this angle at this corner, the 90 degree angle of this square. There's also another angle called the dihedral angle, and for a cube, for example, the dihedral angle is the angle formed between the 2 faces. It's this angle between the 2 objects right here that forms this dihedral angle. If 2 faces meet, this angle and the way they meet, the angle in which they're formed, is the dihedral angle of a polyhedra.

For our polygon we only had 1 notion of angle to worry about, but for polyhedra we have these 2 notions, we have these face angles, the angles on the face themselves, the 4 90 degrees that show up in the polyhedral, and we have these angles between the faces along these edges that show up as the dihedral angles.

Max Dehn, Hilbert's PhD student, solved this problem in a few years after 1900. It was cleaned up with some of the errors corrected and it's now called the Dehn-Hadwiger Theorem. This is a complicated result, but I want to state it roughly as follows.

Here's what it says. If 2 polyhedra P and Q have different kinds of dihedral angles, then they cannot be scissors-congruent, so the dihedral angles will get in the way from them to be scissor congruent or not. In other words, volume is not enough. What do we mean by different kinds of dihedral angles? We care about dihedral angles up to fractional multiples of π.

Let's take a look at 2 examples. These 2 examples are 2 tetrahedra that are in the skeletal structure of the cube. I'm taking my cube, I'm looking at the base of my cube, that square, and I'm cutting the base into 2 pieces. I look at one of those pieces, which is a triangle. That's the base of both of my tetrahedra. They have the same base. Look at the height of this tetrahedra. The one on the left has height, the height of the tetrahedral is placed over the 45 degree angle of this 45, 45, 90 triangle. The one on the right has the same height, it's the same height as the cube for both of them, but it's placed on top of the 90 degree position of the 45, 45, 90 triangle and the base.

They both have the same base, they both have the same height, they both have the same volume, however they are not scissors-congruent. Volume is not enough. Why is this? That is because the Dehn invariant, what Dehn-Hadwiger Theorem says is you want to measure or taste the kind of dihedral angles you have, and if you look at the left tetrahedron, you will see that there are 2 of these edges here and here that have dihedral angle of π over 4, they're 45 degrees. These 3 edges here, here, and here their dihedral angle, the angle formed by those 2 faces meeting, are exactly 90 degrees, π over 2.

This long edge that goes from one corner to the other corner of this cube, the long part of the tetrahedron is π over 3. Notice all of these angles, π over 2, π over 4, π over 3, are all parts of rational multiples of π, fractional parts of π, a half of π, a third of π, a quarter π. But on the right tetrahedron we have these 3 edges here, here, and here that are obviously dihedral angle π over 2. They're 90 degrees because they're exactly parts of the cube themselves. But these 3 edges, the dihedral angles there, turn out to be arc tangent of $\sqrt{2}$.

It is a proof from number theory that arc tangent of $\sqrt{2}$ is not a fractional multiple of π. You can't take π and multiply it by a simple fraction and get arc tangent of $\sqrt{2}$. Since the dihedral angles of these 2 tetrahedra, although they have the same base and they have the same height, are not the same because they have different kinds of dihedral angles. You could never cut 1 up and rearrange it and make it into the other one.

This lecture gives us a taste of discrete geometry focusing on some ideas of proofs and some on intuition. We've built many things from scratch and yet we're able to use all of these by construction. We also defined a new notion of equivalence which turned out to be just area in disguise for polygons; however for polyhedra, the 3-dimensional objects, turn out to be far more complicated than just volume.

Next time we'll push our ideas further in the world of geometry. Stay tuned.

Bending Chains and Folding Origami

Lecture 17

The world of origami can be broken down into 2 worlds, design and foldability. These are the 2 overarching themes when we think of the word origami and the world of origami.

In the last lecture, we focused on cutting and rearranging polygons; in this lecture, we'll look at folding, which has applications in science, technology, and even business. We begin by looking at folding of 1-dimensional linkages, which are modeled by a rod and joint motion. A powerful use of 1-dimensional linkage by this motion is protein folding, which is the process by which a polypeptide chain folds into a 3-dimensional linkage. For proteins, the correct 3-dimensional folded linkage is essential to their function. In fact, several diseases are believed to result from incorrectly folded proteins.

We focus on a much simpler question based on 1-dimensional linkages: When can linkages lock? A linkage is a collection of rods of fixed length in which the joints are allowed to move. A linkage is unlocked if any configuration of a linkage can be deformed and made into any other configuration. A linkage is locked if there are some positions we can get to in the linkage but we cannot move out of to get to any other position.

A linkage in 3 dimensions is almost like a piece of string that can be moved but from a discrete geometric setting. In 1988, it was proven that linkages can lock in 3 dimensions. We see a proof that a 3-dimensional trefoil knot cannot be unlocked. What about 2-dimensional linkages? Given a 2-dimensional linkage with rods of different lengths, can we always unlock the linkage and move the rods around to get a straight line? This is called the Carpenter's Ruler problem, and it was not solved until 2003. The solution showed that there are no locked chains in the plane.

We're all familiar with origami, the centuries-old art of Japanese paper folding. Let's now define the basics of origami folding in a rigorous way. A fold should be something that's isometric in the sense that it preserves distance. In other words, as we fold, we cannot stretch or tear the paper. Further, a paper cannot self-intersect during the foldings; it cannot pass through itself. A fold on a piece of paper is part of a crease line, and creases may come in 2 forms, mountains or valleys. Visually, valleys are notated by dashes and mountains are notated by dash dots.

Origami can be broken down into 2 worlds, design and foldability. Origami design concerns itself with folding a given piece of paper into something having a particular shape. Foldability concerns itself with asking which crease patterns can be folded into an origami pattern that uses exactly the creases on the pattern. And how do we know which patterns will succeed and which ones will not be able to be folded flat without new creases? According to the Maekawa theorem, a necessary (but not sufficient) condition is that the number of mountain and valley folds around each vertex must differ by 2.

… how do we know which ones we can succeed at and which ones will not be able to be folded flat without new creases? There's a classic result called the Maekawa theorem of 1989. It says that the number of mountain and valley folds around each vertex must differ by 2.

A powerful use of origami folding comes in the form of the one-cut conjecture. Think of drawing a polygonal shape on a piece of paper. Is it possible to fold the piece of paper, make one straight cut, and cut out the polygonal shape? The solution depends on lining up all the edges on one line, which of course, is based on folding. ■

Suggested Reading

Devadoss and O'Rourke, *Discrete and Computational Geometry*.

Lang, *Origami Design Secrets*.

Questions to Consider

1. Using a rectangular sheet of paper, make a grid design with random mountain and valley folds. Can you flat-fold this paper? Do this for other random designs.

2. Draw your favorite polygon on a piece of paper and try to cut it out using foldings and one straight cut.

Bending Chains and Folding Origami

Lecture 17—Transcript

Welcome back and thanks for joining me again. The previous lecture was concerned with shapes as seen from the perspective of computational geometry, how somebody in the computer science industry or in the computer science world would work with polygons and manipulate those polygons. They focus on cutting polygons and rearranging them to get other polygons.

This lecture brings the focus on the particular problem of not cutting and gluing, but folding. Folding exists in 1 dimensions. Linkages show up in robot motion planning and in machines. Linkages such as this—where you have balls where the vertices are and joints and rods controlling how these move—show up in, not just robot motion planning, but in protein folding. You can imagine how proteins emerge in certain ways and be folded in different perspectives, which we're going to talk about later. Notice that this is just a 1-dimensional phenomena.

I just have vertices and edges in 3 dimensions that are moving around just like we had knots and links. Now, instead of a continuous perspective of knots and links, I have a discrete geometric perspective where I can count the number of vertices. Notice how the angles change as I move the linkage around. That's folding in 1 dimension, but folding in 2 dimensions also exists.

This is folding paper—origami. For example, leaves of plants rely on origami folding to appear at the right time of year for light absorption. If a leaf is held inside, before it appears, too long then it doesn't get enough light and the plant suffers. On the other hand, if the leaf slowly grows and absorbs light as time goes on it might get too much light for the plant and again, the plant suffers.

What plants do is pack their leaves in certain origami folds so when the right time of year comes, they unfold quickly to absorb exactly the amount of light at the right time. Scientists at Oxford have used origami to create and design new stents. Stents are something we put inside arteries to open up passages. What we want to do is take a small object that's easy to put in and make it into a big object that's able to open up to allow the blood to flow better. These concepts of stent designs have been getting a great influence from origami folding, from traditional methods of folding pieces of paper.

In fact, currently, engineers at MIT have created materials at the nanoscale level, a level that is extremely small. They have then folded these nanoscale materials into 3-dimensional shapes exactly like the way you'd fold paper. Like the brain, with the extra folded layers on the cortex, they believe these nanoscale level foldings will promote faster information flow mimicking the way our brain itself is designed.

Robert Lang, one of the world's top origami designers and artists, has worked with NASA to fold telescopes the size of a football field. Can you imagine a telescope the size of a football field in space, the amount of information it can gather? The big question is, how do you take it there? How do you take this telescope and put it in space? If we can take the telescope, fold it in certain ways, package it into a space shuttle, and unfold it, you can actually transport this. Again, the key is that we don't want to fold it too much because those crease lines, where those foldings occur, is exactly where information is going to be lost.

We want exactly the right kind of folding so that packaging is efficient. In fact, package design in industry is an important cost saving tool for storage and for shipping. If you want to send something over, an empty box for storage, why send the entire box with the empty volume in it? You might as well fold it flat using origami design, send it over, and then unpackage and refold it the way you'd want. You save a lot of storage and you save cost saving values for shipping based on simple things like origami folding.

This subject is far too vast to touch base on every use. Thus, we've just given you a glimpse of how useful it is to think about and understand origami folding—whether it's 1-dimensional folding of linkages or 2-dimensional folding of pieces of paper that motivates these beautiful results.

Let us begin mathematically by looking at folding of 1-dimensional linkages. It is modeled by a rod and joint motion. A powerful use of 1-dimensional linkage by this motion is protein folding. Again, the bar and joint motion is exactly what we talked about. You have a bar or a rod and a joint or the vertex. The vertex edges we had before now have intense flexibility as we move in 3 dimensions. Again, this is what scientists are now using, this method of just vertices and edges, bars and joints, and rods to model proteins.

What is protein folding? Protein folding is the process by which a polypeptide chain folds into a 3-D linkage. You basically have a chain exactly the way we constructed, but as this chain is produced, it starts to become folded. Ribosomes build proteins from the genetic instructions held within the DNA and the RNA. Each protein begins in an unfolded state. A protein, as it's being built, begins as a straight linkage, just a straight bar. As it emerges and comes out of this ribosome state we see that the amino acids interact with this and produce foldings of the protein.

As it starts emerging and as it starts being created, the amino acids bombard this protein and actually start bending it. The resulting 3-D folded protein is determined by the amino acid sequence that bombards it.

For proteins, the correct 3-dimensional folded resulting linkage is essential to its function. Failure to fold into the correct shape produces inactive proteins. In other words, form and function are related once again. The 3-dimensional linkage you get at the end of the day, the 3-dimensional protein that is produced in the way it is created, in fact tells the property of the protein itself. Moreover, several diseases are believed to be resulting exactly from incorrectly folded proteins, such as Alzheimer's, Mad Cow disease, and cystic fibrosis. All of these simply come from proteins that are folded incorrectly.

We see that understanding of proteins is a huge field today from biological, computational, and mathematical perspectives. First, to model it simply with this ball and joint and rod system, is a naïve way of thinking about it. It turns out that that's exactly where we are. That is the forefront of research. This is what the best mathematicians, scientists, and computer scientists are thinking about right now. This field is very young.

Since we cannot dive into protein folding, we focus on a very simple question based on 1-dimensional linkages. We ask the following question: when can linkages lock? A linkage is a collection of rods of fixed length where the joints are allowed to move, exactly the way we talked about before. A linkage is unlocked if any configuration of a linkage can be deformed and made into any other configuration. Think of the analog to knots. In other words, if I'm given a linkage here, it's unlocked if I can make it do anything I want.

On the other hand, a linkage is locked if there're some certain positions you can get to the linkage, but you cannot move out of to get to any other position. It's in a locked state. We study linkages in 2 dimensions, in the plane, and in 3 dimensions in space.

Consider linkages in 3 dimensions. Since this is the closest to a not theoretic field we can have. A linkage in 3-D is almost like a piece of string you can move, but from a discrete geometric setting. Can linkages in 3 dimensions lock? It has been proven by Jason Cantarella and Heather Johnston, in 1998, that linkages can indeed lock in 3-D. The proof is based on the knitting needle example as we see here.

Consider a linkage of 5 segments of length 9, 2, 2, 2, and 9 in this trefoil position you see in front of you. We see that the 6 vertices, these 5 segments form 6 vertices. The joints are going to be labeled 0, 1, 2, 3, 4, and 5. V_0, V_1, all the way up to V_5 could be our 5 vertices. In this particular configuration of these lengths of 9, 2, 2, 2, and 9, I am going to prove to you that this configuration cannot be unlocked. You cannot move this linkage around, this bar, joint, rod system and unlock this thing.

Why can't it be unlocked? Here's the proof. What we do is begin by drawing a ball of radius 4 centered on the midpoint of the middle segment. Remember, you have a segment of 2, 2, and 2. Take the middle 2 segment, take the center of it, take that point, and draw a ball of radius 4 right around that point. By our construction, the points V_1, V_2, V_3, and V_4 are inside this ball during any configuration of our linkage. Those 4 middle points have to be inside it.

Why must they be inside it? Because if it's a ball of radius 4 and if you're picking the middle of that center 1, then that other center part has length 1 and then the extra rod only has length 2, remember it's length 2, 2, 2. You have 1 and a 2, that's only a 3 length. Even if it's stretched out straight, it's only a 3 length. On the other side, you have a 1 and again, a length 2. Stretched out it's only a 3 length; you could never leave a ball of radius 4. No matter what you do with those vertices 1, 2, 3, 4, they're inside this ball.

During any configuration of our linkage, we have this inside our ball since the length of the ending rods are 9—remember, it's 9, 2, 2, 2, 9—the diameter of the entire ball is 8—remember, its radius is 4 so its diameter is 8. But, 9 is bigger than the diameter, right? Nine is even greater than the diameter. Then the points V_0 and V_5 must be outside this ball. Why? Because if V_1 through V_4 are inside, and imagine any points on the inside are V_1 through V_4, they're stuck inside if you put a stick of length 9. Even if V_1 through V_4 are at the extreme end, even if you put a stick of radius 9 it's going to be farther than any diameter of 8. This length of 9 is going to stick out.

Inside are 1, 2, 3 and 4. We know 0 and 5 are outside. Here's what we do. We connect the 0 and the 5 with a very long string. Make sure you keep the string outside of the ball. What do we know? The points are inside—1, 2, 3, and 4—0 and 5 are outside the ball and you have this thing connected by a string which is also outside this ball. If somehow we can unlock this linkage by moving the points in here, inside this ball, then look at what we have. We can make this trefoil knot, which is what you see here in front of you, into the unknot.

If you can unlink it, you can make the trefoil into the unknot. But by tricolorability, from one of the earliest lectures, we proved that the trefoil and the unknot are different. Thus, this linkage can never be unlocked. It must always be in this locked position.

A big unsolved problem currently—which seems quite simple, but again is quite elaborate to even think about in terms of the weapons and machinery needed—is the following: if all our rods are of the same length, can the linkage lock? Remember, here our rods are length 9, 2, 2, 2, and 9. But, what if everything has to have the same length—then can you have any position like this where things lock up? We don't know. We have no idea whether this is true or not.

Let's move on to linkages in 2 dimensions. Can 2-dimensional linkages be unlocked? What does this mean? A 2-dimensional linkage is exactly like a 3-dimensional linkage except I'm going to take my 2-dimensional linkage and put it flat on the table. Given a flat 2-dimensional linkage of different lengths—again the bars and the rods can move on the table—can I always untangle and move these around to get a straight line? Is this possible?

This is called the Carpenter's Ruler problem because a carpenter has a ruler that usually flips open and exactly looks like a linkage, so it's modeled on this Carpenter's Ruler. Let's take a look. Given examples like this or this or this—do any of these examples show positions that are locked? Or is it always possible for us to move the configurations of bars and rods around on the piece of paper, on the plane, to unlock it? Remember, we can move things in 3-D just like knots before. They can't cross each other, but we can move them. In the discrete setting, can I move these linkages and open them up?

This problem was open for 25 years before being solved in 2003 by Robert Connelly, Erik Demaine and Gunter Rote. This turned out to be part of Eric Demaine's PhD thesis at the University of Waterloo. After this, in 2001, just a little bit, he entered MIT as a faculty at the age of 20—the youngest faculty ever. Just a few years after this, in 2003, Eric Demaine was awarded the MacArthur Fellowship, it's called the Genius Award. You are given half a million dollars by the MacArthur Foundation, no questions asked, for being brilliant and for promoting science and society as a whole. They want to make sure money is not an object for you to push with your knowledge.

In future lectures, we are going to see other MacArthur Fellowship winners as well. The theorem states that there are no locked chains in the plane. Eric Demaine's result, along with the others, shows that no matter how you look at chains in the plane, you can always unlock them. This theorem is actually quite difficult and extremely intricate to prove. It is based on something called expansive motions where a method is found, such that the distance between every 2 vertices of the linkage never decreases. In fact, as you're doing this, it seems the distance either stays the same or always increase. If you do this, the linkages unravel and open up beautifully. But, to prove this is beyond the scope of these lectures.

Let us now consider folding—not 1-dimensional linkages, because we're not folding bars anymore—but let's consider folding paper, 2-dimensional pieces of objects. This is called origami. Origami has a rich history. The word origami comes from the Japanese word meaning fold, ori, and paper, gami. Origami itself was probably established with the first creation of paper around 100 A.D. Origami grew in Japan, spreading throughout Japan in the 12th and 13th centuries.

By the 17th century, origami had become a pastime in Japan. The boom in origami in the 20th century is due to Akira Yoshizawa. What Akira Yoshizawa did was he introduced the origami notation of dashes, dots, and arrows in his 1954 book *New Origami Art*. Similar to mathematics, a language is now created. Remember how we came up with a language for tangles and a language for braids and a language for groups? Now we have a language based on his work on origami folding. This enabled people to do powerful things. They were able to translate and transmit information about origami folding that couldn't have been done before.

Instead of talking to someone on the phone, trying to explain what your knot was, now you can give them a braid notation. Instead of talking to someone, explaining what your origami folding is, you can give them this beautiful origami language notation. Consider what we could do with braids, again, once we have this notation.

We want to define the basics of origami folding in a rigorous way. A fold should be something that's isometric in the sense that it preserves distance, iso, the same metric, distance, same distance. In other words, as I fold, I cannot stretch the paper or tear the paper. I have to preserve the distance as I'm folding. A paper cannot also self-intersect during the foldings. It cannot pass through itself, right. I cannot tear the paper. I can only fold it and keep the same distance of what the object was before. I cannot self-intersect the paper, make it go through itself somehow.

A fold on a piece of paper is part of a crease line and creases may come in 2 forms, mountains or valleys. Visually, valleys are notated by dashes and mountains are notated by dash dots. Here we see a simple square piece of paper and I have these dash dots representing this mountain and these 3 dashes representing the valleys. I can take a piece of paper like this, and I can fold it and actually construct exactly this folding pattern, this crease pattern that I have. Just taking your classical double fold of a square piece of paper, look what happens. We have 1 mountain fold—we see this high mountain peak right here—and 3 valleys, they're all the 3 valleys. This fold of this piece of paper is exactly what we see here in this notation.

The entire set of creases form a crease pattern meeting at a common vertex, these endpoints. This vertex here is where these creases meet. We can have complicated transmission of information based on this pattern, like this. Here we have the crane, notice the mountain fold and the valley fold creases. This is the classical crane from the Japanese origami folding. We can also have something like this, which is the base of a frog. This is the frog base. You take this, fold it flat, and use this to build the origami shape of a frog.

The world of origami can be broken down into 2 worlds, design and foldability. These are the 2 overarching themes when we think of the word origami and the world of origami. Origami design concerns itself in folding a given piece of paper into something having a particular shape. This is what most of us think of in terms of origami. A master of origami design, who we've already talked about earlier, is Robert Lang. He has created masterpieces, such as the Black Forest Cuckoo Clock, from one piece of paper.

The ratio of the paper that is needed to fold this object is around 10:1. In other words, it's usually good to use a piece of paper 10 feet long by 1 foot wide. This is what you need to actually build this object and the piece, this Black Forest Cuckoo Clock, involves hundreds and hundreds of folds. His tools, Robert Lang's machines that he uses to do these folding, are his fingers for physical folding. But, he uses geometric and mathematical tools in doing these foldings which people thought were impossible. Math has opened doors to create such beautiful designs that once were impossible because of the language of the fold and the power of mathematics behind it.

Notice, we are starting to blur the lines between art, science, and mathematics. We will come back to this in a later lecture, about how art and science and mathematics indeed fit together. The other world of origami isn't design, it's foldability. Origami foldability concerns itself in asking which crease patterns, the frog base or the crane that we saw before, can be folded into an origami pattern that uses exactly the creases on the pattern.

Which of these objects, given a crease pattern, can you actually fold? We came up with this crease pattern because I actually made it. I actually created this by folding it. What if somebody, a friend of yours or a bitter rival, gives you a crease pattern they designed based on mountains and valleys, just at randomly assigned places? You don't know whether this is a real origami folding pattern or just something they made up, randomly chosen. You're just given it and your job is to make something from this thing, fold it exactly so that the mountains and valleys line up exactly the way the crease pattern tells you to line up. In other words, can it be folded flat? Can I actually take this pattern and fold it the way I want and actually get a flat structure?

We know a mountain right here, we know a mountain, valley, valley, valley fold is possible for this piece of paper. The question is, can I have a mountain, valley, valley fold at this point here? But, what about a mountain, valley, mountain, valley? What if I want this to be a mountain and this to be a valley, but I want this to be a mountain and this to be a valley? What if I want to switch this valley fold into a mountain fold? Is this possible? Let's take a look to see what happens.

If I take this, here's my mountain fold, if I try to make this into a mountain fold also, that's great, here it is. Fantastic! I made 2 mountain folds and 2 valley folds, but I need to fold this flat. How am I going to fold this flat? Notice as I try unsuccessfully to fold it flat, I need to actually make new creases to try to make this folding work, to try to make this flat. These double mountains have constrained me as to what I can do. Remember, I cannot make new creases. These are the only crease information I'm given. You see this is impossible. I cannot do a mountain, valley, mountain, valley.

What about a mountain, mountain, mountain, valley? Is this possible? If you look, it is possible to do a 3 mountain and 1 valley because I do this exact same fold as I did before where I had 1 mountain and 3 valley. I just switch the piece of paper around. Now I have 3 mountains and 1 valley. This is actually possible too and I can flat fold this.

Given some kind of a piece of paper with information like this, how do we know which ones we can succeed it and which ones will not be able to be folded flat without new creases? This is important again and gets to the fundamental core of origami foldability. There's a classic result called the Maekawa theorem of 1989. It says that the number of mountain and valley folds around each vertex must differ by 2.

Let's take a look here. Remember my classic original foldings where I had 1 mountain and 3 valleys. The difference between 3 and 1 is 2. What about the previous examples we looked at earlier? For example for this frog base, if you look at the very central vertex, you see exactly 1, 2, 3, 4, 5, 6 valley folds and 8 mountain folds. The difference between 8and 6 is 2. What about at this corner? Here you see 3 mountain folds and 1 valley fold. The difference between 3 and 1 is 2.

You can check every vertex here in this frog folding pattern or you can check it for the crane and you see it satisfies this theorem. This theorem does not say that this is sufficient. It just says that it's necessary. In other words, just the bare minimum that you need to make sure things work out, at every vertex the difference better be 2. What if, at every vertex, the difference is 2? Does that mean we are guaranteed it works? Not at all. This is the bare minimum you need. You still have to check other conditions.

Do we have an answer to guarantee when a crease pattern can be flat folded? Consider a simple problem like this. It's called the map folding problem, a simpler case of the crease pattern folding. Given a rectangular piece of paper, I'm going to partition and cut apart this rectangular piece of paper by a grid. At each edge of this grid, I'm going to randomly choose a mountain or a valley fold crease such that at every vertex the difference between mountain and valley is 2. You can look at it. It's just a pure grid, vertical and horizontal lines, no complicated folding, just straight classic folding. It's called the map folding problem because this is how you'd fold away your road map.

A quick check will ensure that whether it satisfies Maekawa's theorem—that in fact every vertex does indeed have the difference of 2. Again, this does not guarantee foldability, it just guarantees that this is the first condition you need to check.

It was recently proven that the map folding problem is an extremely hard problem to solve for a computer. In fact, it's proven that it's one of the hardest problems to solve for a computer. In other words, there is no quick check other than, basically, trial and error. This is why it's hard for us to refold road maps. Have you ever gotten frustrated trying to put a road map back together again? That's why. If it's hard for a computer to check through all the possibilities, it's hard for us to go through it one at a time.

I want to close this lecture with a powerful use of origami folding. It comes in the form of the 1 cut conjecture and it states as follows: draw a polygonal shape on a piece of paper. Is it possible to fold the piece of paper, make 1 straight cut, and cut out the polygonal shape? The piece you have in your hand will be on the paper and the drawn polygon will be on the floor. Let's look at some examples just to see what I mean by this.

Here I have a piece of paper with a triangle drawn. Is it possible to fold this piece of paper and make 1 straight cut such that at the end of the day, my triangle is on this table, but the rest of it is untouched in my hand? Let's do this here, take my piece of paper, let's see if I can pull this off live. Remember, what I'm trying to do is try to make sure all the lines of the triangle match up in 1 spot exactly and I get nothing else, but the lines of the triangle. Here I am, I'm going to try to make 1 straight cut and when I do this, in my hand is this sheet of paper and here fallen on the floor is the triangle.

Let's try it for something more complicated. Let's try it for a rectangle, see if we can do this. Let's fold this here and again, I'm trying to match up all my edges, all these edges of these folds on 1 line and exactly 1 line so that nothing else comes here. I do this folding, take my pair of scissors, and making a cut, I open it up. You see, there's the rectangle there inside and this is on the floor. Some of you might be thinking, is it because I'm extremely gifted? The answer is I am, but what about something like this?

Is it possible to make folds so that with 1 cut this complicated ring falls down or what about something that looks like this? Is it possible? Remember, these are all straight line edges to make 1 cut so that this works. This is way beyond my skills. This problem was called, this one cut conjecture. It was originated by Martin Gardner in 1960 in his Scientific American column. The idea has a similar feel to the scissor congruence method of cutting just like we did before. Here we only have 1 cut, but we have the power to fold.

Notice that this depends on lining up all the edges on that 1 line, which, of course, is based on folding. This one cut conjecture became a one cut theorem. It wasn't a conjecture anymore, but it was a theorem itself. It was proven in 2 very different ways and both of which involved Eric Demaine again. He collaborated with other mathematicians and scientists to prove it.

The second method used a disc packing method of how to pack discs inside this polygon that you want to cut, which looks something like this. We have come a long way today, all based on folding. We started with folding and locking 1-dimensional objects in 2 dimensions and 1-dimensional objects in 3 dimensions, these linkages, which were motivated by protein folding. We ended the 1-dimensional section with a Carpenter's Ruler problem. We then folded 2-dimensional pieces of paper due to origami and we ended the 2-dimensional section with a 1 cut theorem.

The next lecture enters the world of polyhedra again and moves from this idea of foldability and flexibility of what we did today into rigidity. Stay tuned.

Cauchy's Rigidity and Connelly's Flexibility

Lecture 18

This results in Arthur Cauchy's rigidity theorem from 1813, one of the most beautiful theorems of polyhedra. It says the following thing: If 2 closed convex polyhedron are combinatorially equivalent—in other words, if they have the same gluing information with congruent faces, the same pieces of the puzzle—then the 2 polyhedra must be identical.

In this lecture, we move from origami and folding of objects to the opposite spectrum, the rigidity of objects. We are motivated by a question of rigidity related to stereoisomers. Recall that stereoisomers are molecules that have the same basic arrangement of atoms and bonds but differ in the way the atoms are arranged in space. There are several different stereoisomers that come from the same set of pieces, the atoms, and their gluing information, the bonds. An example is the dichloroethene molecules, C2H2Cl2, made of carbon, hydrogen, and chlorine. Although they're made up of the same pieces of the puzzle and we're asked to glue the puzzle the same way, the resulting objects are different. And because they are not identical, they have different properties. Form and function are once again related. We'll prove that if stereoisomers form **convex** polyhedra, they must be congruent.

We begin with an understanding of **Cauchy's** rigidity theorem, which states: If 2 closed, convex polyhedra are combinatorially equivalent—in other words, if they have the same gluing information with congruent faces—then the 2 polyhedra must be identical. This theorem has stunning implications for stereoisomers. For a collection of stereoisomers that close up to a convex polyhedral structure, there is only one kind of object that can be made from a chemical perspective. Further, if there is only one way to make this convex object based on the gluing information and the congruent faces, all the dihedral angles of the 2 polyhedra must be the same, which means that they are rigid.

The proof of Cauchy's rigidity theorem is quite ambitious. It is an amazingly complicated result using 2 lemmas and a beautiful method of going back and forth between topology and geometry and converting dihedral angles into angles of polygons. We go through the proof, starting from the assumption that we can find a contradiction to Cauchy's rigidity theorem, but we learn we cannot.

We know that convex polyhedra are rigid, but what about non-convex polyhedra? Can they flex? This remained an open question from the early 1800s until 1976, when Herman Gluck proved that rigid polyhedra are everywhere, although he didn't prove that all polyhedra are rigid. In 1978, Robert Connelly proved that flexible polyhedra exist; he constructed one having 30 triangular faces. Currently, Klaus Steffen has reduced this polyhedron to 14 triangles. This work prompted a new question: As we flex the polyhedron, does its volume change? Does it form a bellows? In the 2-dimensional case, the area changes as the polyhedron flexes, but it has been proven that the volume remains fixed in the 3-dimensional case.

It is … good sometimes to actually look under the hood of some of these theorems, as we did today, to see how mathematicians think and what makes things work.

In the next lecture, we push further forward into geometry by polygons and their uses in terrain reconstruction data. ■

Name to Know

Cauchy, Augustin-Louis (1789–1857): A powerhouse in analysis and a prolific writer, who gave us the arm lemma and the rigidity theorem for polyhedra.

Important Term

convex: An object is convex if the line segment containing any 2 points in the object is contained within the object.

Suggested Reading

Cromwell, *Polyhedra.*

Devadoss and O'Rourke, *Discrete and Computational Geometry.*

Questions to Consider

1. In what other places in nature and in architecture do you see flexible edges but rigid plates?

2. Try to construct a flexible polyhedron yourself. Why is it difficult to do?

Cauchy's Rigidity and Connelly's Flexibility

Lecture 18—Transcript

Welcome back and thanks for joining me again. Today, we move from origami and folding of objects to the opposite spectrum, the rigidity of objects. We are motivated by a question of rigidity in chemistry, about stereoisomers. We talked about this with relation to the symmetry of molecules and the Jones polynomial. We've seen stereoisomers in that context.

What are stereoisomers? As a refresher, stereoisomers are molecules having the same basic arrangement of atoms and bonds, but differ in the way the atoms are arranged in space. There are several different stereoisomers that come from the same set of pieces, the atoms, and their gluing information, the bonds. Although we are told that the atoms and the bonds must be glued and arranged in exactly the same order and fashion, the way they can sit in space can change.

An example is the dichloroethene molecules, $C_2H_2Cl_2$, made of carbon, hydrogen, and chlorine. Take a look. Here, we see 2 examples of this molecule. The one on the left has the chlorine attached to the 2 carbons in the center; there's a chlorine attached to one of the carbons and that carbon is attached to another carbon with a double bond. Notice the carbons are attached to each other and the carbons are attached to the chlorine and the hydrogen on the left.

But, on the right you have the exact same thing as well. You have the 2 carbons in the center attached to each other. Each carbon is attached to a hydrogen and a chlorine, just like the one on the left. But, the one on the left and the one on the right have different structures since they're sitting in space differently. Although they're made up of the same pieces of the puzzle and you're asked to glue the puzzle the same way, the resulting objects are different. Since they are not identical, it turns out they have different properties. Form and function are once again related. We prove below that its stereoisomers form convex polyhedral; they must be congruent. We're going to explore what this means, in a little bit, in more detail. Basically, if stereoisomers form convex polyhedral, there can be only 1 kind of them. You can't have many different versions.

First, we begin with an understanding of an amazing mathematical theorem called Cauchy's Rigidity theorem. We are given a collection of polygonal faces and we are told how each of these faces glue together. Here's the question. How many different resulting objects can we obtain? Remember what we're given again—the same identical pieces of the puzzle and we're told how the puzzle fits together. The question is, how many different answers can you get, at the end of the day, from the pieces and from the gluing information?

It is possible that we might not get a closed polyhedron. Let me explain this a little bit. A polyhedron is closed if it has no boundary. If all the edges have a matching face on either side, there's no boundary to the polyhedron. If the resulting object that we have is not a closed polyhedron, then take a look at this following example. Here we see a polyhedron. It's made up of 4 triangles and it's not closed because I haven't filled in and completed this polyhedron. But, look, I've given the gluing information. This glues to this, this glues to this, and it tells me exactly what the gluing information is. Yet as I flex it, we see that the polyhedron actually changes.

Given the same pieces of the puzzle and the same gluing information, there are numerous ways that I can arrange this object. It fact, it turns out that I have infinitely many different geometric polyhedra I get based on the pieces of the puzzle and how they glue. This one, this one, this one, this one, and this one—these are all different objects. The same pieces of the puzzle give me several different objects.

We can also get only 1 object if the object is a cube with the top of it removed. If we take a cube and remove the top of it, notice that there's only 1 object you can possibly get. Sometimes you will get different objects and sometimes you will end up getting just 1 object if the polyhedron is not closed. Let's pretend our polyhedron now are closed, that it actually completes fully around, that there's no boundary. But, now we assume our polyhedron being closed is not convex. What does this mean?

A polyhedron is convex if any line segment between any 2 points on the polyhedron passes through the polyhedron and not outside of it. Here we see a tetrahedron. Notice, if I pick any 2 points on the polyhedron, the line between them—the site of visibility between those 2 points—passes inside the tetrahedron for any 2 points you could possibly imagine. Thus, the tetrahedron is convex.

What about a cube? A cube is also convex. Any 2 points in the cube can see each other completely inside the cube. What about this object here? It's sort of like a hat of a Viking. Notice that this object is not convex. For these 2 points, you see that that line connecting those 2 points is inside the polyhedron. But, for these 2 points, the line connecting it is outside the polyhedron. Thus, this is not a convex object. We know if the object that we start off with isn't closed, that there could be several different versions of the polyhedron—like the cootie catcher example we just talked about. If the resulting object is not convex, but it does become closed, then again, it turns out that there are several different objects we can get.

Let's take a look. Here, I have the base and sides of a cube. I am going to tell you the gluing information is exactly like this, as the cube says. I use these as the pieces of the puzzle, and I'm actually going to give you more objects to give you a complete closed polyhedron. I can take this cap, made up of 4 triangles, exactly equilateral, and I can put the cap on this polyhedron. I get a closed polyhedron. Notice the green goes with the green, the blue goes with the blue, and the red goes with the red. That's exactly the gluing information I have and it's a closed polyhedron.

But, using the exact same gluing information and the exact same pieces of the puzzle, I can get this polyhedron. Notice, it has the same 4 triangles and it has gluing the same way. The yellow glues with the blue, just like the yellow glued to the blue here, and the green glues with the red, just like the green glues with the red. They also glue here—the yellow with the yellow and the blue with the blue. Notice that I have 2 different possible polyhedron I could obtain based on exactly the same gluing information and the same pieces of the puzzle. We are forced to eliminate these 2 objects first of all. We need to make sure our polyhedron are closed. There's no boundary and we need to make sure our polyhedron are convex, that every point can see every other point. This results in Cauchy's Rigidity theorem from 1813, one of the most beautiful theorems of polyhedra.

It says the following thing: if 2 closed convex polyhedra are combinatorially equivalent—in other words, if they have the same gluing information with congruent faces, the same pieces of the puzzle—then the 2 polyhedra must be identical. Let me say that again. If 2 closed convex polyhedra have the same gluing information with the same pieces of the puzzle, then there's only 1 polyhedron you could possibly build. It must be identical to one another. In other words, this gives the same pieces of the puzzle and the same set of instructions. Only 1 model can be built. As you notice, we saw earlier with a cube with its top pyramid inverted. There are 2 models you could build with the same instructions and the same piece of the puzzle, but yet, it was not convex. The moment you have closed and the moment you have convexity, only 1 piece can be made.

There are some stunning implications of Cauchy's Rigidity theorem. In the world of chemistry, we see that convex stereoisomers—if you have a collection of stereoisomers that close up to a convex polyhedral structure, then we know if we have a convex collection of stereoisomer—must be congruent. In other words, there's only 1 kind of object you can possibly make from a chemical perspective. Thus, you don't need to worry about form and function; there's only 1 thing you can do.

Since there is only 1 way to glue objects to make them convex, what do we notice? If there's only 1 way to possibly make this convex object based on the gluing information and the pieces of the puzzle, that means all the dihedral angles of the 2 polyhedra must be the same. Every angle formed by those 2 edges glued together, that we talked about last time—the dihedral angle that you get between those 2 edges, all of those dihedral angles of the polyhedra—must be the same around every edge. Moreover, since the dihedral angles must be the same, convex polyhedra cannot flex. They must be rigid.

Let's take a look at this example. Here, we have an example of this beautiful structure, the soccer ball design made of a polyhedral version. But, each edge on this polyhedral structure is flexible. I can actually flex along this edge. It's made up exactly like the pieces that I showed you earlier. But, the moment I make it convex and the moment I close it up so there are no holes, it completely forms the shape of a completed polygonal sphere. This cannot flex. Although each 1 has freedom to move, together there is no freedom at all, it must be rigid. Since it did flex, if somehow this did flex, then I would have another polyhedron which was not congruent to this one. It would have the same pieces of the puzzle, because it would be just made up of the same thing just flexing, and it would be having the same gluing information. But Cauchy's Rigidity theorem says that you can only have 1 such thing. You can't have 2 objects that have the same gluing information and the same pieces of the puzzle which give you convexity and closure at the same time.

The proof of Cauchy's Rigidity theorem is broken into 3 steps. I'm going to give you a warning. This is one of the most ambitious ventures we're going to undertake. These 3 steps are easy to follow one by one. But, at the end, when we culminate and put it together again, it involves almost every piece of the puzzle we have used so far in making sense of what we're doing. Thus, I encourage you to look through and see this again if the first time around it didn't make full and complete sense.

Step one of Cauchy's Rigidity theorem's proof involves a 1-dimensional linkage theorem, a linkage idea that we talked about last time, called Cauchy's Arm Lemma. This could be easily called Cauchy's Arm theorem. A lemma from a mathematical perspective is a true result, just like a theorem is a true result. The only reasons mathematicians use lemma instead of theorem is for a subjective reason. We believe lemmas are not as great as theorems. If you think a lemma is a great result, then you have the right to call it a theorem. That's the only difference. Moreover, these lemmas that we're going to talk about—the step 1 and the step 2, where we're going to build these lemmas—are going to be used to prove Cauchy's Rigidity theorem. Thus, we call it a lemma because it's not as important as the end result.

What does this Cauchy's Arm Lemma say? Let's take a look. If a chain, a convex chain as you see here on the plane, has some or all of its internal angles increase then the distance between its 2 endpoints must increase. Look at this convex chain. It's convex because any point can see any other point inside this collection, inside this closed polygon you get. You have this chain. If you pick some of the internal angles here, here, and here for example, and you increase them—not too much, not over 180 degrees, not over being flat, but if you just increase them—then the distance between the 2 endpoints has to also increase.

This seems almost obvious. Yet, the proof comes from the fact that the cosine function that you see here—the cosine X, the X-axis and the Y-axis is the height of the cosine—is an increasing function between negative π to 0. As you see from negative π to 0, cosine is an increasing function. The proof of this Arm Lemma is based on that result. Although we don't want to talk about it in detail now, I just want to give you a glimpse of what the geometric implications are. That's the first step. The first step was the Cauchy's Arm Lemma.

The second step involves a simple Two Coloring Lemma. Again, we call this a lemma because we're going to use it later and we don't think it's as important as this theorem coming up. Let G be any connected graph on the plane made up of vertices and edges, any connected graph you can draw on the plane that is arbitrarily colored with 2 colors. You draw any graph on the plane and you choose to color the edges either red or blue. It's your choice, any way you want for any graph you want.

Here's an example of a graph I drew. I've colored the edges arbitrarily red and blue. What does this Two Coloring Lemma say? It says, as we walk around each vertex, let's count the number of times the colors change. If you look at this vertex, notice it goes from red to blue and then blue to blue, blue to red, and red to red again. There's 2 color changes. It goes from red to blue once and then it goes from the blue to red. There's 2 color changes. This vertex over here has 4 color changes. It goes from blue to red and then it goes from, as you spin around, it goes from red to blue and then blue to red and then red to blue again. There are 4 times the color change. Do this for different vertices.

The lemma says, no matter what graph is drawn on the plane and no matter how it's colored—taking red and blue—there will always be a vertex with at most 2 color changes. At most, you'll only have 2 color changes for any vertex as you walk around it. Notice here, we have that vertex with 2 color changes. We have several vertices which only have 2 color changes as we walk around. In this particular case, the lemma is satisfied. But this is true for any graph you can imagine and any color use of desire. There's always a vertex with at most 2 color changes. That's what this lemma says. Notice how this graph breaks the plane into vertices, edges, and faces. You see all those vertices; we can count 19 vertices for this graph. We can count 22 edges and there's 4 closed regions—those are my faces—plus my outside big face, that's my fifth face. You see we have Euler's formula again—vertices, 19, minus edges, 22, plus the number of faces, 5, equals 2. Remember, the Euler formula $V - E + F = 2$. The proof of this lemma, the 2 coloring lemma, uses Euler's formula to prove it.

Again, we don't want to go into the details of this proof right now because we're interested in Cauchy's Rigidity theorem. Given these 2 lemmas as facts—we're going to assume that they've already been proven—I want to show you the proof of Cauchy's Rigidity theorem. It involves an elegant mixing of the previous 2 lemmas, the Arm Lemma and the Two Coloring Lemma. Here's what I'm going to do. Let P and Q be 2 polyhedra that are combinatorially equivalent—they're made up of the same pieces of the puzzle glued the same way—but, let's pretend they have different dihedral angles. We're going to assume that we have the same pieces of the puzzle glued the same way, but they turn out to be 2 different ones. We're going to assume that Cauchy's Rigidity theorem is wrong and we're going to find a contradiction. We must find a contradiction that this cannot happen.

We're going to pretend we have 2 different polyhedral, P and Q, that have the same pieces of the puzzle with the same gluing instruction. But, somehow, the dihedral angles are different. One is bent differently than the other. Here's what I'm going to do. I'm going to color the edges of P—so, take my polygon P and look at the edges of P—blue if its dihedral angle is more than that of Q. Remember, this edge, there's a corresponding edge over here. Each edge has another edge, each vertex has another vertex, and each face has matching faces. Remember, it's made up of the same pieces of the puzzle glued the same way. For every face, edge, and vertex, there's another guy over here.

I'm going to take one of my edges, look at P alone, I'm going to color this blue if its dihedral angles are more than Q. I'm going to look at some edges and I'm going to color them red if those dihedral angles here happen to be less than Q. Many of those other edges, if they have the same dihedral angles, I'm going to leave it alone.

I know some of my edges are going to be colored differently than Q because of the fact that they're different because the dihedral angles are different. If they had identical dihedral angles, they'd be the same. But, I'm assuming they're different dihedral angles. I color them blue, I color them red, and those that are the same, I leave alone. We are now going to convert this geometric problem about angles into a topological problem first. Remember that we can lay flat polyhedral graphs of P by removing 1 its faces. Putting my fingers in, I'm going to stretch it and put it flat on the plane. Let me use that trick. I'm going to take my polyhedron P. I'm going to take a face, stretch it, and lay it flat, topologically. I've lost all angle information, but I just want to know which is touching which.

Looking at only the colored edges, since it's flat on the plane—looking at only the colored edges and forgetting all the other ones—I choose a connected graph from these. Maybe there're a bunch of colored edges over here on this plane. Maybe there're a bunch of colored edges over here. I just look at one of the set of colored edges that's connected. Now, what do I have? I have a graph made up of vertices and edges, these colored edges, on the plane. By the previous Two Coloring Lemma, there is a vertex somewhere on here with at most 2 color changes, by our step 2 above, our second lemma.

This vertex, on this plane, corresponds to a vertex on my polyhedron P. Now that we have found our special vertex to focus on, on our polyhedron, we enter the land of geometry again. We have used topology to do what we want and we enter geometry. What do we do now? We take this vertex here. Remember, the moment we have a vertex in this polyhedron, we have another vertex, the identical vertex in this polyhedron. I'm going to take a sphere of radius, really small value, extremely small value. I'm just going to put a small sphere of a small radius around that vertex. I'm going to put a small sphere of the same radius around the vertex over here corresponding to Q.

This cuts out a spherical polygon around P and it cuts out a spherical polygon around Q. Remember, if I take this sphere and if I intersect it—if I intersect a sphere with my polyhedron—I would get some polygon, some polygonal piece over here. Note that a dihedral angle of the polyhedron, this angle of this polyhedron, now gets converted to a polygonal angle, to that polygon's angle. I've now converted a 2-dimensional dihedral angle problem between 2 faces to a 1-dimensional angle problem between 2 edges. Now we are beginning to see where Cauchy's polygonal Arm Lemma could fit in.

We know that this vertex that I picked, V, has either 0 or 2 color changes for the edges incident of V. It only has 0 or 2 color changes. Remember, that's the special vertex I picked the first time by laying it flat. But, a color change of an edge incident to V means a change in the dihedral angle of the edge of the polyhedron. Remember, that's what those color changes meant in the first place. If it's blue, the angle is bigger and if it's red, the angle is smaller. Let's consider the 2 possible cases of 0 or 2 color changes separately. I'm looking at V, I'm looking at around V, that small region, and I get this polygon around this.

If no sign changes are around V—if it has 0 color changes—then all the colored edges of the polygon P touching V must be of 1 color. All of them must be blue or all of them must be red. Let's just say, for the sake of argument, that they're all blue. Thus, all angles of the polygon from P are either larger, if the edge is blue, or equal to, if it has no color, than the polygon from Q. We get a polygon that we cut out from the sphere intersecting this. We look at the polygon. Since the dihedral angles are bigger—because they're all blue over here—than Q, that means my polygon over here must have bigger angles than the one from Q.

Delete an edge in both of the corresponding polygons. Delete the corresponding edge from both of these polygons, 1 from the polygon P and 1 from the polygon Q. Since the angles in P are different, this deleted edge must have different links for P and for Q. Let me explain why. Let me cut that edge up. Since the angles of P are bigger than those of Q—remember, because these have those blue edges—then that edge connecting them by Cauchy's Arm Lemma says that extra length that you get closing up that chain must be a longer 1 than it is for Q.

But, this edge length is actually just the length of a part of the face of the polyhedron. Remember, we took the sphere, cut it, and we got that edge length right there. Thus, my edge lengths couldn't change because I started with the same pieces of the puzzle, all faces are congruent. This is impossible. I cannot have a vertex with 0 color changes around it. But, I know that my vertex V, that I picked, has either 0 color changes or 2 color changes. I just eliminated the 0 color change possibility or else the polygon around here would have been bigger than this. By Cauchy's Arm Lemma, that's impossible.

What if there are 2 color changes? Let's take a look. If there are 2 color changes, we get a situation like this. We use a similar type of idea as earlier. If 2 color changes, then it must go from a blue part—maybe there's a bunch of blue points which correspond to these vertices increasing, these dihedral angles increasing—to a collection of red points. Maybe that corresponds to them decreasing. Remember, there're only 2 color changes. Thus, you can have a bunch of blues going to red and maybe a bunch of reds going to blue—only 2 color changes.

Here's what I'm going to do. I'm going to draw an edge between these 2 transitional parts between the red/blue part and the blue/red part. Consider the length of this edge from the blue side of the polygon, from the blue's perspective. Blue says, things are getting bigger, the dihedral angles are getting bigger—which means for this edge, length from the Q side it must increase. Look at it from the red side of the polygon's perspective. Red says, things are getting smaller, which means the dihedral angles are decreasing. These angles are getting smaller, so this edge length must have gotten smaller.

Cauchy's Arm Lemma says, this edge must have different angle lengths, 1 smaller and 1 bigger. But, this is impossible; you can't have an edge with 2 possible length changes, smaller and bigger. Thus, this vertex cannot have 2 color changes. But, we know that the vertex V, from the very beginning, has either 2 or 0 color changes. Yet we show both of these cases cannot happen.

What does that mean? There's a contradiction. We assumed something incorrectly at the very beginning. We assumed some of the dihedral angles of P were different than Q. Our very original step, when we started coloring things red and blue because they were different, was where we messed up. That was a wrong assumption. Thus, all the dihedral angles are the same and the polyhedra are identical. That is Cauchy's Rigidity theorem. It is an amazingly complicated result using these 2 lemmas and a beautifully intricate way—going into topology using Euler's formula, jumping back into geometry, talking about converting dihedral angles into angles of polygons. Ah, it's gorgeous.

I have skipped certain parts throughout this proof, which can be detailed later on. I encourage you to look at a bigger picture setting of how this proof works. It is actually good sometimes to actually look under the hood of some of these theorems, as we did today, to see how mathematicians think and what makes things work.

What do we know? We know convex polyhedra are rigid. But, what about non-convex polyhedral? We know that if the polyhedra is closed in convex, it has to be rigid. But, what do we know about non-convex polyhedral? Can non-convex polyhedra flex? If we go back to our original case over here, notice that this is the non-convex polyhedra of going from this to this. But, this is not really flexing. I can't flex this and make it into this; I can only replace these pieces and go like this. They are these 2 independent possible possibilities. The question is, can you have a non-convex polyhedron that actually flexes? This was an enormous open problem since 1813. Herman Gluck, in 1976, proved that in the space of all polyhedra—if you have a space where each point in the space is a different polyhedral—the rigid polyhedra turn out to be everywhere like the rational numbers in the real number line. The rigid polyhedra are just everywhere; everywhere your eyes look, there are rigid polyhedra.

He didn't prove that all polyhedra were rigid. But, he said the rigid ones are everywhere. Have a fun time trying to find something that flexes. The belief was that there is not going to be any polyhedron, convex or non-convex, that flexes. In fact, if you go home and try to build these on your own, you will see that it is quite difficult to make anything that flexes at all. You have to close it up with all the edges being able to flex. But, at the end of the day, it must close up and you'll notice it becomes rigid.

In a stunning result, Robert Connelly, in 1978, proved that there exist flexible polyhedra. He constructed this with 1 having 30 triangular faces. He was motivated by a flexible object in 4 dimensions, which Connelly then tried to project into 3 dimensions to obtain this inspiration. We're going to talk about these 4-dimensional objects in future lectures.

Currently, Klaus Steffen has reduced this polyhedron with 30 triangles to a polyhedron of only 14 triangles. Here is a picture of this polyhedron constructed by my students, Rohan Mehra and Norm Nicholson, made of Plexiglas in a course I taught at Williams. Notice how it actually flexes along these edges. Every edge is hinged; the whole polyhedron becomes not convex and yet it flexes. Because of this, a new question came. As we flex the polyhedral, does its volume change? Does it form a bellows? You know how you have a bellows to pump air into the fireplace. Does the polyhedra form a mathematical bellows? If you poke a hole in it and if you flex it, does the volume change? Is the air getting pumped in and out of this polyhedron?

Let's take a look. For the 2-dimensional case, if I have a polygon and if I can flex along these vertices—these correspond to flexing along edges in 3-D—then look. Here's my area and as I flex, my area is decreasing or it could increase or decrease. In fact, I can make my area as small as I want. In the 2-dimensional version, it's certainly the case that area changes as we flex. For the 3-D case, Dennis Sullivan conjectured that the volume actually remains fixed. In 1997, Connelly and Walz, along with this monumental work by Sabitov, proved that this is true.

Volume remains fixed as you flex this polyhedron. Beautiful work! They proved that no matter what polyhedron you have—convex or non-convex, rigid or flexible—volume will always be the same if it can flex or if it can't. An amazing powerful geometric result has been shown today—Cauchy's Rigidity theorem. It is an extremely ambitious theorem to prove. I wanted to show you under the hood to see this engine involving geometry of polygons, Euler characteristics in disguise, and the cleverness of putting it all together elegantly.

In our next lecture, we push forward more into geometry by polygons and their uses in terrain reconstruction data. Stay tuned.

Glossary

amphicheiral: An object is amphicheiral if it can be made into its mirror image.

configuration space: The space that keeps track of all possible ways an object (or a collection of objects) can be arranged.

convex: An object is convex if the line segment containing any two points in the object is contained within the object.

convex hull: Given a point cloud, its convex hull is the smallest convex set containing the point cloud.

dimension: An invariant given to a point on a shape that measures the degrees of freedom afforded at that point.

Fields Medal: The highest honor given for mathematical research; recipients must be under 40 years of age.

fractal: An object that holds a high level of self-similarity.

group: An algebraic structure given to a collection of elements with a means of combining the elements (composition) satisfying three conditions (identity, inverse, associativity).

homeomorphic: A notion of equivalence, weaker than isotopic. Two objects are homeomorphic if one object can be cut up into pieces, stretched, and reattached along the cuts to form the other object.

homotopic: A notion of equivalence, weaker than homeomorphic. This notion deals only with continuous deformations where self-intersections are allowed.

isotopic: A notion of equivalence, the strongest in the world of topology. Two objects are isotopic if they differ by stretching (rubber sheet geometry).

knot: A circle placed in three dimensions without self-intersections.

manifold: A generalization of a surface to higher dimensions, where each point on the manifold has a neighborhood having the same dimension.

phylogenetic tree: A mathematical tree structure that shows the relationship between species believed to have a common ancestor.

polytope: The higher-dimension version of a polygon and a polyhedron.

Schlegel diagram: A diagram of a polytope that allows it to be depicted using one less dimension.

scissors-congruent: A notion of equivalence. Two objects are scissors-congruent if one can be cut up and rearranged into the other.

surface: An object on which every point has a neighborhood that has two degrees of freedom.

surgery: The process of cutting and regluing 3-manifolds.

Biographical Notes

Cauchy, Augustin-Louis (1789–1857): A powerhouse in analysis and a prolific writer, who gave us the arm lemma and the rigidity theorem for polyhedra.

Conway, John H. (1937–): Conway is a professor at Princeton and a prolific mathematician whose works encompass geometry, group theory, number theory, and algebra. In particular, he is known for his Conway notation for knots and links.

Dehn, Max (1878–1952): A student of David Hilbert, Dehn is known for his work in geometry and topology, particularly Dehn invariants for scissors-congruence of polyhedra and Dehn surgery for manipulating 3-manifolds.

Demaine, Erik (1981–): Demaine is an expert in computational geometry who became a professor at MIT at the age of 20. He has solved numerous unsolved problems, such as the one-cut theorem and the Carpenter's Rule problem, becoming a MacArthur Fellow in 2003.

Euler, Leonhard (1707–1783): One of the greatest mathematicians of all time, his scientific works cover analysis, number theory, geometry, and physics. He was one of the first to use topology, from which we receive the formula $v - e + f = 2$ of a polyhedron.

Fejes-Tóth, László (1915–2005): One of the fathers of modern discrete geometry, his works influence practically all areas of this field today. In particular, he investigated packings and partitions and laid the framework for understanding the Kepler conjecture, later solved by Hales.

Gauss, Carl Friedrich (1777–1855): Known as the Prince of Mathematics, Gauss is considered by many to be the greatest mathematician since antiquity. His foundational work in all areas of mathematics continues to influence our world today. We get the notion of curvature and the powerful Gauss-Bonnet theorem from him.

Hales, Thomas (1958–): Hale solved two powerful open problems (some of the oldest ones in discrete geometry): the Kepler conjecture and the honeycomb conjecture. He believes that partnership with computers will be fundamental in solving future mathematical problems.

Jones, Vaughan (1952–): Winner of the Fields Medal in 1990, he created one of the most powerful knot invariants.

Kelvin, William Thomson, Lord (1824–1927): A powerful scientist, Kelvin had wonderful notions of shape and nature. He believed that knots embodied properties of atoms and worked with soap bubbles to posit an efficient tiling of space. He is best known for his Kelvin temperature scale of absolute zero.

Kepler, Johannes (1571–1630): A mathematician and astronomer, Kepler tried to relate platonic solids to the solar system. He also made a conjecture about the best way to stack spheres in space. He is most well known for discovering the elliptical motions of planets around the Sun.

Klein, Felix (1849–1925): Klein spearheaded some of the pioneering relationships between algebra and topology. He also showed us how to obtain all surfaces from gluing polygons.

Lang, Robert (1961–): Lang is not only a world-class origami artist, but he is also a leader in the field of mathematical and computational origami, designing and folding previously unimaginable shapes using mathematics.

Mehretu, Julie (1970–): An artist who focuses on large-scale paintings depicting motion, space, movement, and high-level historical references, Mehretu constructs her work with layers of acrylic paint on canvas. She became a MacArthur Fellow in 2005 and has had exhibitions all over the world, including at the Museum of Modern Art and the Williams College Museum of Art.

Perelman, Grigori (1966–): Perelman completed the work of Richard Hamilton, using Ricci curvature flows to prove the Poincaré conjecture and, most likely, Thurston's geometrization conjecture itself. Although he won the Fields Medal in 2006, he did not accept it.

Poincaré, Henri (1854–1912): One of the greatest and most prolific mathematicians in history, Poincaré worked in geometry, algebra, number theory, physics, and the philosophy of science. He is credited with being the father of modern topology.

Riemann, Bernhard (1826–1866): A student of the great Gauss, Riemann revolutionized the study of shapes by separating topology and geometry into two worlds with the brilliant notion of a metric.

Thurston, William (1946–): A pioneer in the field of topology in dimension 3, Thurston gave us the geometrization conjecture, describing the possible geometries of all 3-manifolds. He was awarded the Fields Medal in 1982.

Weeks, Jeffrey (1956–): A student of Bill Thurston's, Weeks uses his understanding of 3-manifolds to study the shape of the universe. He became a MacArthur Fellow in 1999 and wrote the program *Curved Spaces*.

Witten, Edward (1951–): A mathematical powerhouse who received the Fields Medal in 1990, Witten is considered the greatest physicist of our time, known for his work in string theory.

Bibliography

Adams, Colin. *The Knot Book.* Providence, RI: American Mathematical Society, 2004. This is one of the first and simply the best book on knots and links. Written at an elementary level, it provides numerous details on the subject, ranging from elementary notions of knots, to surfaces, to 3-manifolds.

Aste, Tomaso, and Denis Weaire. *The Pursuit of Perfect Packing.* 2nd ed. Boca Raton, FL: Taylor and Francis, 2008. Written in the form of short chapters touching on numerous points in science and nature, this book addresses such issues as the Kepler conjecture, the Kelvin cell, and the Weaire-Phelan structure. It includes beautiful examples from physics, biology, chemistry, and engineering dealing with packing and partitioning.

Cromwell, Peter. *Polyhedra.* New York: Cambridge University Press, 1999. This is probably the most accessible source for the understanding of polyhedra in all aspects. It has beautiful illustrations and covers numerous topics, such as regularity, rigidity, Gauss-Bonnet, and colorings.

Devadoss, Satyan, and Joseph O'Rourke. *Discrete and Computational Geometry.* Princeton, NJ: Princeton University Press, in press. The only book of its kind, this text brings a topic from computer science into the realm of mathematics. It is a beginning college-level book that discusses such ideas as scissors-congruence, triangulations, Voronoi diagrams, and convex hulls from a geometric viewpoint.

Devaney, Robert. *A First Course in Chaotic Dynamical Systems.* Boulder, CO: Westview Press, 1992. An advanced undergraduate book on chaos and dynamics with lots of details; presented in a quite readable fashion.

Hatcher, Allen. *Algebraic Topology.* New York: Cambridge University Press, 2001. This graduate-level book is one of my favorites. It is elegantly written, with clean and clear illustrations (for a graduate-level math book). It shows algebraic topology from a very visual perspective.

Lang, Robert. *Origami Design Secrets*. Natick, MA: AK Peters, 2003. This book is the magnum opus of the world's greatest origami expert. Although not written at an elementary level, it is well organized, beginning with simple techniques and moving toward deep and powerful tools for folding and design. It has the strongest mathematical underpinnings of any origami book and provides designs for some the greatest creations ever produced in this art form.

O'Shea, Donal. *The Poincaré Conjecture*. New York: Walker and Company, 2007. A mathematical sweep of history is seen through the lens of the Poincaré conjecture in this book, written for non-experts. It tracks the geometric and topological struggles and (most importantly) failures of this great problem, bringing us to the great solution by Hamilton and Perelman.

Richeson, David. *Euler's Gem*. Princeton, NJ: Princeton University Press, 2008. In this book, we see the world of topology and geometry through the eyes of Euler's formula and its generalizations. It's a fun read that shows the power of mathematics weaved into a historical framework.

Tufte, Edward. *Envisioning Information*. Cheshire, CT: Graphics Press, 1990. In one of my all-time favorite books, Tufte helps us to see beautiful and efficient ways of displaying data. This book is accessible and stunning to behold; it serves as a launching point in struggling with the visualization and depiction of shapes.

Weeks, Jeffrey. *The Shape of Space*. 2nd ed. Boca Raton, FL: CRC Press, 2001. A beautifully written, easy-to-read book on building and understanding 3-manifolds. It deals with the geometry and the topology of the universe in clear and knowledgeable terms.

Wilson, Robin. *Four Colors Suffice*. Princeton, NJ: Princeton University Press, 2004. This beautifully written book focuses on the history and mathematics of the four-color theorem. It is written at an elementary level, accessible to all.

Notes

Notes